METEOROLOGY AND ENERGY SECURITY

Simulations, Projections, and Management

METEOROLOGY AND ENERGY SECURITY

Simulations, Projections, and Management

Edited by
Paul S. Samuel, PhD

Apple Academic Press Inc.	Apple Academic Press Inc.
3333 Mistwell Crescent	9 Spinnaker Way
Oakville, ON L6L 0A2	Waretown, NJ 08758
Canada	USA

©2016 by Apple Academic Press, Inc.

First issued in paperback 2021

Exclusive worldwide distribution by CRC Press, a member of Taylor & Francis Group

No claim to original U.S. Government works

ISBN 13: 978-1-77463-709-8 (pbk)
ISBN 13: 978-1-77188-386-3 (hbk)

Library and Archives Canada Cataloguing in Publication

Meteorology and energy security : simulations, projections, and management/edited by Paul S. Samuel, PhD.

Includes bibliographical references and index.
Issued in print and electronic formats.
ISBN 978-1-77188-386-3 (hardcover).--ISBN 978-1-77188-387-0 (pdf)
1. Meteorology. 2. Power resources. 3. Energy security. I. Samuel, Paul S., editor
QC861.3.M48 2016   551.5   C2015-908308-7   C2015-908309-5

Library of Congress Cataloging-in-Publication Data

Names: Samuel, Paul S.
Title: Meteorology and energy security : simulations, projections, and management / [edited by] Paul S. Samuel, PhD.
Description: Toronto : Apple Academic Press, 2016. | Includes bibliographical references and index.
Identifiers: LCCN 2015047106 (print) | LCCN 2015049332 (ebook) | ISBN 9781771883863 (hardcover : alk. paper) | ISBN 9781771883870 ()
Subjects: LCSH: Meteorology. | Weather forecasting. | Power resources. | Energy security. | Energy consumption--Forecasting.
Classification: LCC QC861.3 .M48 2016 (print) | LCC QC861.3 (ebook) | DDC 333.79/13--dc23
LC record available at http://lccn.loc.gov/2015047106

Apple Academic Press also publishes its books in a variety of electronic formats. Some content that appears in print may not be available in electronic format. For information about Apple Academic Press products, visit our website at **www.appleacademicpress.com** and the CRC Press website at **www.crcpress.com**

About the Editor

PAUL S. SAMUEL, PhD

Paul S. Samuel earned his PhD in mechanical engineering from the University of Manitoba, Winnipeg, Canada, in 2007. He was a post-doctoral fellow/research associate at the Department of Mechanical and Manufacturing Engineering, University of Manitoba, and the Department of Fisheries and Oceans (DFO), Winnipeg, Manitoba, Canada, during 2007–2008. His research interests cover computational and numerical turbulent flows around bluff bodies, turbulent flow over smooth and rough surfaces, andalso renewable energy, with special focus on wind energy potential in severaldeveloping countries. He has over 35 technical publications and hasheld lecturing positions at the Obafemi Awolowo University, Ile-Ife, Nigeria.

Dr. Paul is currently a professional engineer in Manitoba with Sunrit Engineering Consulting Services, Inc.

Contents

Acknowledgment and How to Cite

The editor and publisher thank each of the authors who contributed to this book. The chapters in this book were previously published in various places in various formats. To cite the work contained in this book and to view the individual permissions, please refer to the citation at the beginning of each chapter. Each chapter was read individually and carefully selected by the editor; the result is a book that provides a nuanced look at meterology and energy security. The chapters included are broken into five sections, which describe the following topics:

- Chapter 1 presents useful TMYs for 35 Chinese cities, which can be used as input data for predicting solar energy production.
- Chapter 2 discusses accurate TMYs, useful for building energy simulations, that will lay the groundwork for future energy efficiency study.
- Chapter 3 contains important data on how weather and wind energy converter system components interact, which can be used in order to carry out detailed assessments of potential wind farms.
- Chapter 4 gives research on the marine atmospheric boundary layer, which is important to offshore wind farms.
- Chapter 5 contains a parameterization to represent wind turbines in weather prediction models in order to predict capacity density and wind resource limitations.
- Chapter 6 summarizes a potential problem with wind farms: very large installations may change the regional weather, resulting in less wind (and less energy production) than had been anticipated.
- Chapter 7 uses modeling, observational data, remote sensing, and geographic information systems to calculate solar radiation more accurately.
- Chapter 8 discusses the effect that global climate change, as a result of greenhouse gas emissions, will have on sunlight predictions.
- Chapter 9 discusses the vulnerability of Ghana's hydropower to weather patterns and climate change.
- Chapter 10 explains the importance of accurate weather forecasts for achieving more effective and efficient hydropower operations.
- Chapter 11 offers probability methodology for estimating summer and winter peak hours demands.

- Chapter 12 offers an energy management technique that will compensate for wind and solar energy's lack of dependability and power quality.
- Chapter 13 considers the future implications for the electricity grid of large-scale dependence on solar and wind energy.

List of Contributors

A.S. Adams
Department of Geography and Earth Sciences, University of North Carolina at Charlotte, Charlotte, NC 28223-0001, USA

Fidelis I. Abam
Department of Mechanical Engineering, Michael Okpara University of Agriculture, Umudike, Abia State, Nigeria

Muyiwa S. Adaramola
Department of Ecology and Natural Resources Management, The Norwegian University of Life Sciences, Ås, Norway

Iftikhar Ahmed
Baluchistan University, Quetta, Pakistan

Ismail Hakki Altas
Department of Electrical and Electronics Engineering, Engineering Faculty, Karadeniz Technical University, 61080 Trabzon, Turkey

Emmanuel Obeng Bekoe
CSIR-Water Research Institute, Achimota, Accra, Ghana

P. Block
International Research Institute for Climate and Society, Columbia University, Palisades, New York, USA

Pengwei Du
Pacific Northwest National Laboratory, P.O. Box 999, Richland, WA 99352, USA

Richard L. Fagbenle
Department of Mechanical Engineering, Obafemi Awolowo University, Ile-Ife, Nigeria

B.H. Fiedler
School of Meteorology, University of Oklahoma, Norman, OK 73072-7307, USA

Evangelia-Maria Giannakopoulou
EDF Energy R&D UK Centre, 52 Grosvenor Gardens, London SW1W 0AU, UK

Huei-Ping Huang
School for Engineering of Matter, Transport and Energy, Arizona State University, Tempe, AZ 85281, USA

Katsuhiro Ichiyanagi
Department of Electrical and Electronics Engineering, Aichi Institute of Technology, 1247 Yachigusa, Yakusa-cho, Toyota, Aichi 470-0392, Japan

Hans E. Jørgensen
Wind Energy Department, Technical University of Denmark, Risø Campus, Frederiksborgvej 399, Roskilde, Denmark

Fredrick Yaw Logah
CSIR-Water Research Institute, Achimota, Accra, Ghana

Jakob Mann
Wind Energy Department, Technical University of Denmark, Risø Campus, Frederiksborgvej 399, Roskilde, Denmark

D. McMillan
Wind Energy Systems Centre for Doctoral Training, 204 George Street, University of Strathclyde, Glasgow, G1 1XW, UK

Onur Ozdal Mengi
Department of Energy Systems Engineering, Engineering Faculty, Giresun University, 28100 Giresun, Turkey

F. M. Mulder
Faculty of Applied Sciences, Materials for Energy Conversion and Storage, Delft University of Technology, P.O. Box 5045, 2600GA Delft, The Netherlands

Regis Nhili
EDF Energy R&D UK Centre, 52 Grosvenor Gardens, London SW1W 0AU, UK

Olayinka S. Ohunakin
Mechanical Engineering Department, Covenant University, Ota, Ogun State, Nigeria

Olanrewaju M. Oyewola
Department of Mechanical Engineering, University of Ibadan, Oyo State, Nigeria

Erik Lundtang Petersen
Wind Energy Department, Technical University of Denmark, Risø Campus, Frederiksborgvej 399, Roskilde, Denmark

Gerardo Andres Saenz
School for Engineering of Matter, Transport and Energy, Arizona State University, Tempe, AZ 85281, USA

Shahzad Sultan
Institute of Space and Earth Information Science, Chinese University of Hong Kong, Hong Kong, China.

Ib Troen
Wind Energy Department, Technical University of Denmark, Risø Campus, Frederiksborgvej 399, Roskilde, Denmark

G. Wilson
Wind Energy Systems Centre for Doctoral Training, 204 George Street, University of Strathclyde, Glasgow, G1 1XW, UK

Renguang Wu
Institute of Space and Earth Information Science, Chinese University of Hong Kong, Hong Kong, China;Shenzhen Research Institute, Chinese University of Hong Kong, Hong Kong, China.

Qingshan Xu
School of Electrical Engineering, Southeast University, Nanjing 210096, China

Haixiang Zang
School of Electrical Engineering, Southeast University, Nanjing 210096, China

Jay Zarnikau
Frontier Associates LLC, Austin, USA and LBJ School of Public Affairs and Division of Statistics and Scientific Computing, The University of Texas at Austin, Austin, USA

Shuangshuang Zhu
Frontier Associates LLC, Austin, USA

Introduction

PART 1: TYPICAL METEOROLOGICAL YEAR GENERATION

A typical meteorological year (TMY) is a collation of selected weather data for a specific location, generated from a data bank collected over the space of years. The data is selected to cover a range of weather phenomena for the location in question, while still giving annual averages that are consistent with the long-term averages for the location in question.

This information has become increasingly important in a world that is trying to wean itself off from its oil dependency by turning to other "clean" energy sources. TMY data provides the essential data that can be used in building simulation, in order to assess the expected heating and cooling costs for the design of the building. It is also used by designers of solar energy systems, from solar hot water systems for homes to large-scale solar thermal power plants.

In chapter 1, the authors propose modifying the concept of a typical meteorological year (TMY) by using measureable weather data—dry bulb temperature, relative humidity, wind velocity, atmospheric pressure, and daily global solar radiation—to generate a TMY database. They use a total of eleven weather indices to create TMYs for 35 Chinese cities during 1994–2010. The authors then take their research a step further, and make a practical connection between TMY and solar energy production.

In chapter 2 we shift our attention from China to Nigeria, where the authors use 23-year hourly weather data—including global solar radiation, dew point temperature, mean temperature, maximum temperature, minimum temperature, relative humidity, and wind speed—to create a TMY for the region. After comparing TMY predictions with the 23-year long-term average values, they found that the close agreement indicates these TMYs can be used in building an energy simulation study, laying the groundwork for future energy efficiency.

PART 2: WIND ENERGY AND THE WEATHER

Wind energy is actually a form of solar energy, since winds are caused by the sun's uneven heating of the atmosphere (combined with the irregularities of the earth's surface and its rotation). Terrain, bodies of water, and plant cover all modify wind flow, which can then be harvested by wind turbines to generate electricity. Many factors need to be considered when building a wind farm, but the primary consideration is meteorological data to determine the most efficient locations. (These locations may then need to be ruled out or modified because of other factors such as population location and bird migration patterns.)

In our third chapter, data is collected from two large-scale wind farm sites in order to create a model that calculates wind speed dependant failure rates—in other words, technical problems caused by weather variations. This information is then used to for simulation for three potential UK wind farm sites. The authors' results show that the control system and the yaw system are the components most affected by changes in average daily wind speed. The model takes into account the effects of the wind speed on the cost of operation and maintenance and also includes the impact of longer periods of downtime in the winter months and shorter periods in the summer. Understanding these factors allows for a detailed site assessment, which has significant practical value for operators and engineers.

The marine atmospheric boundary layer (MABL) is very important to offshore wind farms, since atmospheric stability has a direct relation to wind and turbulence profiles. A better understanding of the stability conditions occurring offshore and of the interaction between MABL and wind turbines is needed, which in chapter 4 the authors obtained in a study of atmospheric conditions over the North Sea.

As referenced earlier when discussing chapter 2, planning and financing wind farms requires accurate resource estimation (in addition to a number of other considerations relating to environment and economy.) The authors of chapter 5 remind us that there is another important factor: the spacing of turbines for maximum power production. This can have a Catch-22 affect, in that very large wind farms may actually change regional wind patterns, resulting in the geophysical limit to wind power production often being lower than it had been predicted to be. The authors cite work believable data that indicates that when wind energy has been implemented on a very massive scale, it will affect the power production from entire regions. This factor has to be taken into account when considering wind farm planning.

Chapter 6 gives subgrid parameters that represent wind turbines in weather prediction models. Researching the problem presented in chapter 5, the authors parameterization models the turbines' drag and atmospheric mixing, as well as the electrical power production the wind causes in the turbines. Next, the authors apply their model to a study of wind resource limits in a hypothetical giant wind farm in order to better understand capacity density.

PART 3: SOLAR ENERGY AND THE WEATHER

Solar power converts the sun's energy into thermal or electrical energy. It is the cleanest and most abundant renewable energy source available. Engineering technology harnesses this energy for various uses, including generating electricity, providing light, heating and cooling homes and other buildings, and heating water for domestic, commercial, or industrial use. As the world turns to this alternative form of energy for building our future energy security, the industry is working to scale up the production of solar technology, and drive down manufacturing and installation costs. Understanding and predicting weather patterns is a vital part of this.

The authors of chapter 7 investigate the spatial distribution of direct solar irradiation on rugged terrain over Pakistan. They used a digital elevation model (DEM), along with observational data of solar radiation from six meteorological observatories in Pakistan and five Chinese meteorological observatories. The emerging field of remote sensing and geographic information systems (GIS) makes it possible to calculate solar radiation with finer spatial resolution and better understanding of terrain and cloud effects. The authors' findings will be useful for both countries as they further their solar energy development.

One of the essential factors that must be considered when projecting solar energy potential is the effect of global climate change. Sunlight patterns will change. The authors of Chapter 8 discuss the climate-change model for the USA, which shows significant regional sunlight changes. Despite this, the authors conclude that the estimates of solar power potential using present-day climatology will remain useful in the coming decades.

PART 4: HYDROPOWER AND THE WEATHER

Hydropower is energy taken from moving water and converted to electricity. The most common method of energy capture is a hydroelectric dam, where water coming down through an area causes turbines to rotate, and the energy is captured to run a generator. Power can also be generated from the energy of tidal forces or waves.

Hydropower is a key energy source in almost all regions of the world. (Canada has the highest use, while the USA comes in second.) Hydropower fuels social and economic development, ensures electricity security, and is an essential component of our future energy security.

There are environmental considerations that affect hydroplant construction and planning, but overall, hydropower is a practical energy alternative. It is relatively cheap, and it is more reliable than sunlight and wind. However, weather can also play a role in hydropower, and must not be forgotten. Of even greater concern is the fact that hydropower is vulnerable to climate change.

Ghana's electricity comes mainly from hydropower generated from a dam at Akosombo built in 1965. Because of harsh weather conditions, such as flooding and droughts, the country has frequently suffered from serious electric power rationing. Power rationing meant that factories and cities could not function, and foreign investment in Ghana's economy was jeopardized. Chapter 10 analyzes 37 years of rainfall in the Volta basin and intake water levels in the dam site on the Volta Lake for hydropower generation to establish whether the main causes of the power rationing was actually due to drought. The authors determine that in most cases, weather was the culprit. The authors report the economic consequences. The also suggest that climate change will bring further negative effects to the country's water resources. They call for sustainable management, so that drought and floods will not negatively impact hydropower generation in future.

Seasonal precipitation forecasts are seldom integrated with water resources operations. Planning ahead for "bad weather" is practically non-existent, even in regions of scarcity. The author of chapter 11 attributes this to human nature, which would rather plan for the best-case scenario and not think about the risks. Convincing demonstrations of forecast value could change this, since inaccurate meteorological predictions encourage people to discount weather predictions.

PART 5: SEASONAL ENERGY MANAGEMENT

As any homeowner knows, weather extremes are the number-one cause of high bills. Only a few days of extreme temperatures can increase power usage and cause household energy bills to be noticeably higher. This same principle applies on a larger scale. Our future energy security depends on accurate assessments of what our actual needs will be, based on what the weather will be.

In chapter 12, the authors propose a probability-based method to estimate the summer and winter peak demand reduction from an energy efficiency measure when TMY data and model simulations are used to estimate peak impacts. In the estimation of winter peak demand impacts from some example energy efficiency measures in Texas, their proposed method performs better than two alternative methods. It doesn't fare quite as well in its estimation of summer peak demand impacts, but the results are still useful.

Since the wind and solar sources are not reliable in terms of dependability and power quality, a management system is required for supplying the load power demand. In chapter 13, the authors introduce an intelligent energy management system for maintaining the energy sustainability

If we are going to depend on solar and wind energy for our future energy security, we will need large-scale implementation of solar and wind-powered electricity generation that will use continent-sized electricity grids built to distribute the locally fluctuating power. Systematic power output variation will be apparent, since both sunlight and wind fluctuate on a daily basis. Coping with such constant variations in power generation will require large-scale renewable energy storage and storage capacities for at least up to half a day. Seasonal timescales for energy conversion and storage will be equally necessary. The authors of chapter 15 have created a first-order model for estimating required energy storage and conversion magnitudes, taking into account potential diurnal and seasonal energy demand and generation patterns.

—Samuel S. Paul

PART I

TYPICAL METEOROLOGICAL YEAR GENERATION

A Modified Method to Generate Typical Meteorological Years from the Long-Term Weather Database

HAIXIANG ZANG, QINGSHAN XU, PENGWEI DU, AND KATSUHIRO ICHIYANAGI

1.1 INTRODUCTION

China lies in the northeast part of East Asia between 4° and 53° North latitude and 73–135° East longitude with a population of about 1.3 billion [1, 2]. China, as the largest developing country, is the second largest country in energy consumption [3]. To relieve the dual pressure from rising energy demand and growing environmental problems, renewable energy sources are considered for satisfying a significant part of the energy demand in China [4–6]. As one of the renewable energy, solar energy is a clean energy source and is extremely abundant in China. More than two-thirds of China receives an annual total solar radiation above $5.9\,GJ/m^2$ ($1639\,kWh/m^2$) with more than $2200\,h$ of sunshine per year [7–9]. On the other hand, solar radiation is the fuel of solar energy systems. Solar radiation data are the basic and key parameters in the applications of solar energy systems [9–11].

A Modified Method to Generate Typical Meteorological Years from the Long-Term Weather Database. © *Zang H, Xu Q, Du P, and Ichiyanagi K.* International Journal of Photoenergy **2012** *(2012), http://dx.doi.org/10.1155/2012/538279. Licensed under Creative Commons Attribution 3.0 Unported License, http://creativecommons.org/licenses/by/3.0.*

Since solar radiation data can vary from year to year, there is a need to generate a customized solar radiation database that can well represent the long-term averaged solar radiation over a year [12, 13]. Representative databases for one year duration, known as typical meteorological year (TMY) [14], are often employed for computer simulations of solar energy conversion systems and building systems. Based on the TMY method, typical solar radiation data are formed by the selection from the real recorded weather data.

In the past, several methodologies [15–20] for forming TMYs have been reported, such as Sandia method, Festa-Ratto method, and Danish method. Among the different, TMY generation methods, the Sandia method is widely adopted. Hall et al. [19], Said and Kadry [21], Marion and Urban [22], Petrakis et al. [23], Kalogirou [24], Chow et al. [25], Wilcox and Marion [26], Yang et al. [27], and Jiang [14] generated TMYs for different locations with different weather parameters and assigned weighting factors. These methods in above literatures are in fact similar, the main differences lie in the numbers of daily indices (weather indices) to be included and their assigned weightings [25].

A few studies for selecting TMYs in China have been found in recent years. In the paper authored by Chow et al. [25], typical weather year files for Hong Kong and Macau were produced and analyzed. Zhou et al. [9] developed typical solar radiation years and typical solar radiation data for 30 meteorological stations in China only using the long-term daily global solar radiation records. In the paper of Yang et al. [27], TMYs for 60 cities in five major climatic zones of China were investigated through nine recorded weather indices (only the global solar radiation and not including the direct solar radiation data). But the data of the nine measured weather indices are before the year of 2000. Jiang [14] generated TMYs using nine weather parameters (including the global and the direct solar radiation data). However, only eight cities were considered and generated.

It is suggested that the TMY selection process should include the most recent meteorological observations [25]. In this present study, based on the modified TMY method (eleven meteorological indices and the novel assigned weighting factors), the TMYs and typical solar radiation data for 35 stations are formed and analyzed using the latest and accurate long-term weather data.

TABLE 1: Geographical locations and data period of the meteorological stations used for this study.

Number	Location	Latitude (N)	Longitude (E)	Elevation (m)	Period	Total years
1	Beijing	39°48'	116°28'	31.3	1994–2010	17
2	Changchun	43°54'	125°13'	236.8	1994–2010	17
3	Changsha	28°13'	112°55'	68	1994–2010	17
4	Chengdu	30°40'	104°01'	506.1	1994–2003	10
5	Dongsheng	39°50'	109°59'	1460.4	1994–2010	17
6	Fuzhou	26°05'	119°17'	84	1994–2010	17
7	Guangzhou	23°10'	113°20'	41	1994–2010	17
8	Guiyang	26°35'	106°44'	1223.8	1994–2010	17
9	Haikou	20°02'	110°21'	13.9	1994–2010	17
10	Hami	42°49'	93°31'	737.2	1994–2010	17
11	Hangzhou	30°14'	120°10'	41.7	1994–2010	17
12	Harbin	45°45'	126°46'	142.3	1994–2010	17
13	Hefei	31°52'	117°14'	27.9	1994–2010	17
14	Jiamusi	46°49'	130°17'	81.2	1994–2010	17
15	Jinan	36°36'	117°03'	170.3	1994–2010	17
16	Kashgar	39°28'	75°59'	1288.7	1994–2010	17
17	Kunming	25°01'	102°41'	1892.4	1994–2010	17
18	Lanzhou	36°03'	103°53'	1517.2	1994–2003	10
19	Lhasa	29°40'	91°08'	3648.7	1994–2010	17
20	Mohe	53°28'	122°31'	433	1997–2010	14
21	Nagqu	31°29'	92°04'	4507	1994–2010	17
22	Nanchang	28°36'	115°55'	46.7	1994–2010	17
23	Nanjing	32°00'	118°48'	7.1	1994–2010	17
24	Nanning	22°38'	108°13'	121.6	1994–2010	17
25	Sanya	18°14'	109°31'	5.9	1994–2010	17
26	Shanghai	31°24'	121°29'	6	1994–2010	17
27	Shenyang	41°44'	123°27'	44.7	1994–2010	17
28	Taiyuan	37°47'	112°33'	778.3	1994–2010	17
29	Tianjin	39°05'	117°04'	2.5	1994–2010	17
30	Urumqi	43°47'	87°39'	935	1994–2010	17
31	Wuhan	30°37'	114°08'	23.1	1994–2010	17
32	Xian	34°18'	108°56'	397.5	1994–2004	11
33	Xining	36°43'	101°45'	2295.2	1994–2010	17
34	Yinchuan	38°29'	106°13'	1111.4	1994–2010	17
35	Zhengzhou	34°43'	113°39'	110.4	1994–2010	17

1.2 SELECTION OF CITIES AND DATA USED

In order to organize and generate the TMY database, the daily weather data are required. The available weather data are managed and provided by China meteorological stations. Due to space limitation, 35 cities having the local meteorological stations are investigated and selected. These stations cover latitudes range from 18°14′N (Sanya) to 53°28′N (Mohe), longitudes from 75°59′E (Kashgar) to 130°17′E (Jiamusi) and have considerably varied altitude from 2.5 m (Tianjin) to 4507 m (Nagqu). All the complex climates within China are represented in the cities, and the relevant information for the 35 stations is shown in Table 1.

In view of the actual situation in China and the characteristics of solar energy systems, eleven meteorological indices are applied in this paper. These weather indices are maximum, minimum, and mean dry-bulb temperature; minimum and mean relative humidity; maximum and mean wind velocity; maximum, minimum, and mean atmospheric pressure; daily global solar radiation. In addition, the relative errors of global solar radiation recorded data are changed from ±10% to ±0.5% since 1993 owing to new observation instrument in China meteorological stations. So, the most recent and accurate weather data during the periods between 1994 and 2010 are chosen and gathered for the present research.

The missing and invalid measurements, accounting for 0.38% of the whole database, are marked and coded as 32744 or 32766 in the data. Moreover, these problematical data are replaced by the values of previous or subsequent days using the interpolation method [15]. In the data processing, if more than 5 days measured data are not available in a month, the month is eliminated from the database.

1.3 METHODS USED

The Typical Meteorological Year (TMY) method, developed by Sandia National Laboratories, is selected and modified in this paper [19]. It is one of the most widely adopted methodologies for combining 12 typical meteorological months (TMMs) from different years over the available period to form a complete year. The procedure for selecting the 12 TMMs consisted of two steps.

1.3.1 SELECTION OF FIVE CANDIDATE YEARS

According to the Finkelstein-Schafer statistic [28], if a number, n, of observations of a weather index x are available and have been sorted into an increasing order x_1, x_2,...,x_n, the cumulative distribution function (CDF) for this weather index is determined by a monotonic increasing function CDF(x). The formula of function CDF(x) is given as follows:

$$CDF(x) = \begin{cases} 0 & \text{for } x < x_1, \\ \dfrac{(i - 0.5)}{n} & \text{for } x_i \leq x \leq x_{i+1}, \\ 1 & \text{for } x \geq x_n. \end{cases}$$

(1)

The FS comparison statistics between the long-term CDF for each month and the CDF for each individual year of the month are calculated by the following equation:

$$FS_x(y, m) = \frac{1}{N} \sum_{i=1}^{N} |CDF_m(x_i) - CDF_{y,m}(x_i)|$$

(2)

where $FS_x(y,m)$ is FS(y,m) statistics for each weather index x (y means year and m means month); CDF_m is the long-term and $CDF_{y,m}$ is the short-term (for the year y) cumulative distribution function of the weather index x for month m; N is the number of daily readings of the month (e.g., for January, N = 31).

Based on the FS statistic and the meteorological database used in this work, the CDF curves of daily global solar radiation (DGSR) for the months of June and December (choosing Nanjing as an example) are shown in Figures 1 and 2. It can be concluded that the short-term CDFs follow quite closely their long-term counterparts. In Figure 1, the CDF of DGSR for June 2000 is most similar to the long-term CDF (smallest value of FS statistic), while the CDF of DGSR for June 2005 is least similar (largest value of FS statistic). Likewise, in Figure 2, the DGSR CDF for December of 2006 is closest to the long-term CDF, while the DGSR CDF for December of 2010 is most dissimilar. Even though they are not the best months, respect to the long-term CDF, June of 2007 and December of 1998 are finally selected as the TMM for June and December, respectively. This is a consequence of additional selection steps described in the following.

For each of the candidate months, the FS statistics of the eleven weather indices are grouped into a composite weighted sum (WS) by (3). Moreover, the five years with the smallest WS values are selected as the candidate years

$$WS(y, m) = \frac{1}{M} \sum_{x=1}^{M} WF_x \cdot FS_x(y, m)$$

(3)

where WS(y,m) is the average weighted sum for the month m in the year y; WF_x is the novel weighting factor for the xth weather index (in Table 2); M is the number of meteorological indices (11 in this study).

The novel assigned weighting factors, which are significant for generating typical meteorological data, are shown in Table 2. A large weighting factor of 0.5 is assigned to global solar radiation because the criteria is mainly used for solar energy systems and the other weather variables (e.g., dry bulb temperature, relative humidity, and atmospheric pressure) are affected by solar radiation. For instance, in general, the higher for the solar radiation, the higher for the dry-bulb temperature. Furthermore, the weighting factor of the mean weather index is larger than that of the max and min weather parameter, for example, 2/24 for Mean Relative Humidity and 1/24 for Min Relative Humidity.

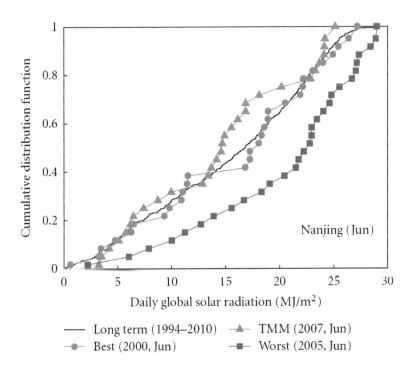

FIGURE 1: CDFs for June daily global solar radiation for Nanjing.

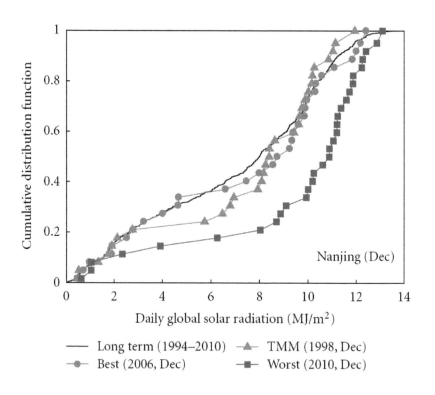

FIGURE 2: CDFs for December daily global solar radiation for Nanjing.

TABLE 2: Meteorological indices and weighting factors for the FS statistic in this study.

Number	Parameter indices		Present article
1	Temperature	Max dry bulb temperature	1/24
2		Min dry bulb temperature	1/24
3		Mean dry bulb temperature	2/24
4	Humidity	Min relative humidity	1/24
5		Mean relative humidity	2/24
6	Wind	Max wind velocity	1/24
7		Mean wind velocity	2/24
8	Pressure	Max atmospheric pressure	1/48
9		Min atmospheric pressure	1/48
10		Mean atmospheric pressure	1/24
11	Solar radiation	Global solar radiation	12/24

1.3.2 FINAL SELECTION OF TMM

The final selection of the TMM from the five candidate years involved a selection process by Pissimanis et al. [29], simpler than the original Sandia method. This method utilizes the root mean square difference (RMSD) of global solar radiation

$$RMSD = \left[\frac{\sum_{i=1}^{N}\left(H_{y,m,i} - H_{ma}\right)^2}{N} \right]^{1/2}$$

(4)

where RMSD is the root mean square difference of global solar radiation; $H_{y,m,i}$ is the daily global solar radiation values of the year y, month m, and day i; H_{ma} is mean values of the long-term global solar radiation for the month m; N is the number of daily readings of the month.

Moreover, the month with the minimum RMSD is finally selected as the TMM.

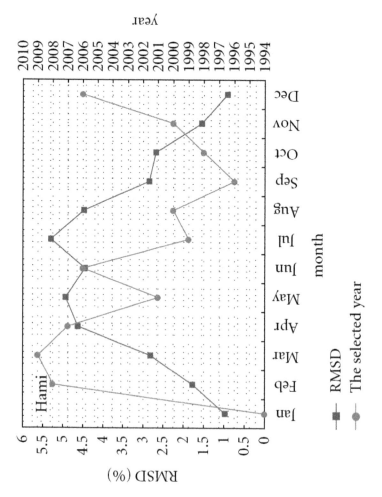

FIGURE 3: The minimum RMSD and the selected years for each month of the year.

TABLE 3: Summary of the TMYs for the 35 stations in China.

Number	Station	Month											
		Jan	Feb	Mar	Apr	May	Jun	Jul	Aug	Sep	Oct	Nov	Dec
1	Beijing	2007	1998	2004	1996	2010	1997	2001	2002	2000	2004	2004	2000
2	Changchun	2009	1997	2006	2007	1996	2001	2003	2008	2006	2006	2006	1999
3	Changsha	2004	1997	2007	2007	2004	2003	2008	2000	1998	2003	1999	2006
4	Chengdu	1998	1998	1996	2001	2002	2000	2000	2002	2000	2000	1999	1994
5	Dongsheng	1997	2006	2000	2007	1997	2006	2004	2005	2000	1998	2002	1999
6	Fuzhou	1999	2006	1995	1997	2002	1995	1998	2008	1998	2007	1999	1998
7	Guangzhou	1999	2003	2003	2006	2001	2007	2008	2003	1998	1999	1999	2004
8	Guiyang	1995	2002	2005	2008	2001	2007	2000	2007	2001	2007	2004	2005
9	Haikou	1994	1994	2001	1998	2004	2003	1998	1999	2000	1996	2005	1998
10	Hami	1994	2008	2009	2007	2001	2006	1999	2000	1996	1998	2000	2006
11	Hangzhou	2005	1994	2004	1999	1996	2000	2000	2008	2001	1997	2005	1998
12	Harbin	2003	1998	2005	1996	2001	1995	1998	2009	1996	2004	2009	1996
13	Hefei	2003	2006	2004	1997	2005	2003	2008	1996	2002	1994	1999	2009
14	Jiamusi	2009	1997	2000	2008	2001	1995	2010	2008	2002	2008	2009	2003
15	Jinan	1997	1997	2010	2007	2008	2010	2007	2001	1999	2005	2001	2006
16	Kashgar	2005	2004	2006	1998	1997	2005	2001	2003	2000	1998	1998	1997
17	Kunming	2003	1998	2001	2005	1995	2004	2000	1997	2005	2000	1998	2000
18	Lanzhou	2000	1997	2000	2000	2003	1998	2002	2000	2002	1998	1999	2003
19	Lhasa	2001	1996	2006	2008	1994	1994	1996	2005	2001	2000	2001	2001
20	Mohe	2010	1999	2006	2010	2010	1999	2006	2006	2007	1998	2005	2006

TABLE 3: *Cont.*

Number	Station	Month											
		Jan	Feb	Mar	Apr	May	Jun	Jul	Aug	Sep	Oct	Nov	Dec
21	Nagqu	1995	2003	1995	1996	2000	1999	1998	2009	2008	1995	1998	1994
22	Nanchang	2010	1995	2004	1994	1998	2007	2008	2004	1996	2003	1999	1998
23	Nanjing	1997	1995	1995	2004	1997	2007	1995	1996	1996	1996	2001	1998
24	Nanning	2004	2002	2005	2000	2002	2000	2009	2004	2004	2010	2003	2004
25	Sanya	2002	2002	2001	2003	1998	2000	2000	1999	2004	1995	2002	1996
26	Shanghai	1997	1997	2004	2008	2000	1998	1998	2004	2003	1997	1999	1998
27	Shenyang	2009	1997	2004	2007	2006	1996	2005	2002	1995	2006	2006	1999
28	Taiyuan	1996	1995	2006	2006	2005	1998	1995	2009	2006	1999	2001	2003
29	Tianjin	2009	1999	2000	2004	2002	1997	2005	2002	1999	1999	2004	1999
30	Urumqi	1999	1996	2007	2004	2005	2006	1994	1995	2005	2008	2006	1996
31	Wuhan	1996	2000	2009	2001	1997	2003	2005	1995	2004	2008	1999	1998
32	Xian	1996	1996	1999	1995	1997	2003	2000	1999	1999	1996	2004	1998
33	Xining	2000	2001	2010	2000	2005	2009	2007	2002	2006	2008	2007	1997
34	Yinchuan	1998	2003	2000	2003	1997	2006	2001	1997	1999	2003	1999	2003
35	Zhengzhou	1997	1998	1995	1995	1997	1998	2002	1995	2000	2008	2001	1998

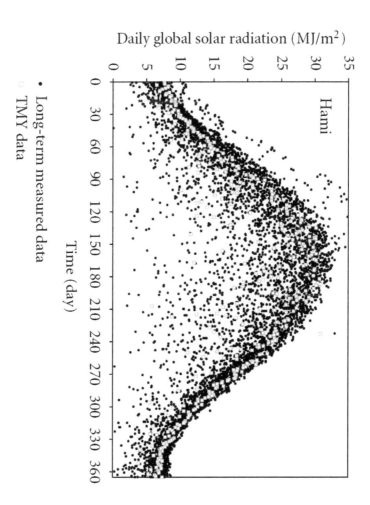

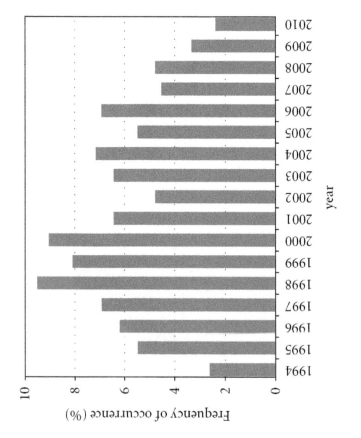

FIGURE 5: Summary of year selection frequency for 1994–2010 TMY.

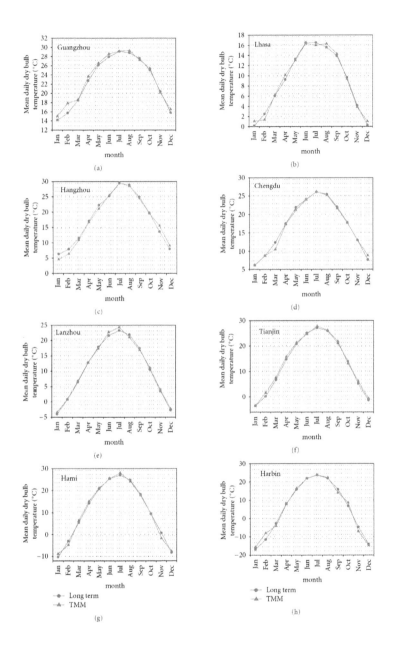

FIGURE 6: Monthly mean daily dry bulb temperature for the long-term and for the selected TMMs at eight stations.

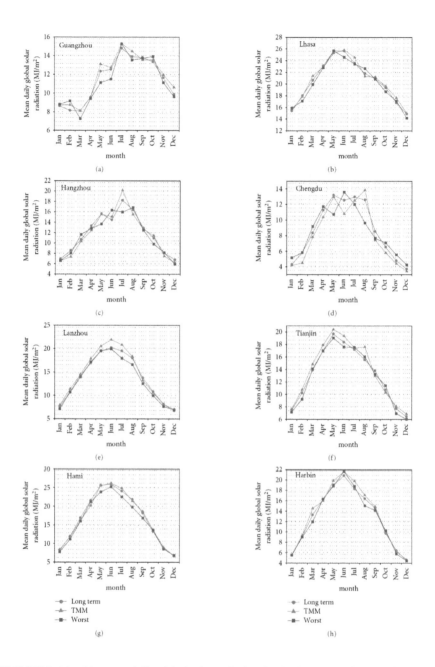

FIGURE 7: Monthly mean daily global solar radiation for the long-term, for the selected TMMs, and for the worst months at eight stations.

1.4 RESULTS AND DISCUSSION

Using the above modified TMY method, the typical meteorological years (TMYs) for the 35 stations listed in Table 1 are generated, consisting of the selected most typical years for 12 months. Furthermore, the detailed typical solar radiation data sourced from the TMYs are analyzed in the following.

Table 3 shows a summary of the TMYs selected for the 35 stations in China. In the final selection, the month with the minimum RMSD of global solar radiation is selected as the TMM. The selected months for creating the TMY of Hami and the minimum RMSD in the corresponding month are shown in Figure 3. From Figure 3, the value of the RMSD of global solar radiation varies between 0.92% (Dec of 2006) and 5.30% (Jul of 1999). And it is obviously that several years are used to form the typical months, and data of year 2000 and 2006 are more used than other years in the TMY of Hami.

Although typical solar radiation data obtained from the TMYs are formed for all the 35 stations in the research, owing to space limitations, it is not practical to present all of them in the paper. So, only typical solar radiation data of the Hami station are shown in Figure 4. These data would be useful for the designers of solar energy systems in China. In Figure 4, the maximum and minimum values of daily global radiation data are 30.80 MJ/m^2 on July 6 and 3.87 MJ/m^2 on December 18, respectively. Figure 4 shows the variation of typical solar radiation data derived from the TMY data and the long-term (1994–2010) measured data for Hami stations. It can be seen that both the long-term and the typical solar radiation data are variable and fluctuant through the year. Furthermore, typical solar radiation data (green panes) and the long-term recorded data are comparable in size.

In order to find out which years tend to follow the long-term weather patterns more closely than the others, the selected TMYs listed in Table 3 are investigated. The year selection frequency for the TMYs derived from the period of 1994–2010 is shown in Figure 5. In Figure 5, it is apparently that the frequency occurrence of the year 1998 is up to 9.52% and the year 2010 is only 2.38%. For intensive study, Table 4 gives a summary of the times which the year is selected as the TMM. Obviously, 1998 and 2010 are the most frequent year (40 times) and the least frequent year

(10 times), respectively. In each year during the period 1994–2010, the number of the times vary from one month to another and 10 (November of 1999) is the largest selected number. In other words, no more than 10 stations are selected for the same month in any particular year.

TABLE 4: Frequency of a year being selected to provide a TMM.

Year	Jan	Feb	Mar	Apr	May	Jun	Jul	Aug	Sep	Oct	Nov	Dec	Total times
1994	2	2	0	1	1	1	1	0	0	1	0	2	11
1995	2	3	4	2	1	3	2	3	1	2	0	0	23
1996	3	3	1	3	2	1	1	2	4	3	0	3	26
1997	5	7	0	2	7	2	0	2	0	2	0	2	29
1998	2	5	0	2	2	4	5	0	3	5	3	9	40
1999	3	2	1	1	0	2	1	3	4	3	10	4	34
2000	2	1	5	3	2	4	6	3	6	3	1	2	38
2001	1	1	3	3	5	1	3	1	3	0	5	1	27
2002	1	3	0	0	4	0	2	5	3	0	2	0	20
2003	3	3	1	2	1	5	1	2	1	3	1	4	27
2004	2	1	6	3	2	1	1	3	3	2	4	2	30
2005	2	0	3	1	4	1	3	2	2	1	3	1	23
2006	0	3	5	2	1	4	1	1	3	2	3	4	29
2007	1	0	2	5	0	4	2	1	1	2	1	0	19
2008	0	1	0	4	1	0	4	4	1	5	0	0	20
2009	4	0	2	0	0	1	1	3	0	0	2	1	14
2010	2	0	2	1	2	1	1	0	0	1	0	0	10

In addition, the accuracy of TMY data is excellent on monthly bases. Figures 6 and 7 show the monthly mean values of the actual dry bulb temperature and global solar radiation for eight stations (Guangzhou, Lhasa, Hangzhou, Chengdu, Lanzhou, Tianjin, Hami, Harbin), respectively. There are three monthly profiles in Figures 6 and 7, which are determined by the long-term (1994–2010), the selected TMMs, and the worst months (the month with the largest RMSD value in the final selection of TMM). From Figures 6 and 7, it can be concluded that the monthly mean daily dry bulb temperature and global solar radiation derived from the TMY data represent good agreement with the long-term recorded average data. Moreover, in Figure 7, in general, the difference between the long-term measured data with the TMY data is smaller than that of the worst months.

1.5 CONCLUSIONS

Typical solar radiation data are the key elements for the applications of solar energy systems and building energy simulation. A modified TMY method using the Finkelstein-Schafer statistical to generate TMY data from the long-term measured weather database is implemented at 35 stations in China. In particular, eleven weather indices and the novel assigned weighting factors are proposed and applied in this research.

TMYs for 35 stations in China are investigated and generated based on the accurate and the most recent long-term (1994–2010) measured weather data, such as dry bulb temperature, relative humidity, wind velocity, atmospheric pressure, and daily global solar radiation. Typical solar radiation data obtained from the TMY data, with Hami as an example, are presented and analyzed in this study. It is found that the global solar radiation cumulative distribution functions (CDFs) of the selected TMMs tend to follow their long-term counterparts well. It is also seen that the year 1998 follows the long-term weather patterns more closely than the others. Moreover, there is a good agreement between the typical solar radiation data and the long-term measured data on monthly basis.

It is expected that the TMY data and typical soar radiation data generated in this paper will exert positive effects on some energy-related scientific researches and engineering applications in China. Future researches will focus on the TMY data and typical solar radiation data on a larger regional scale. We hope to report on these works in the near future.

REFERENCES

1. L. Q. Liu, Z. X. Wang, H. Q. Zhang, and Y. C. Xue, "Solar energy development in China-a review," Renewable and Sustainable Energy Reviews, vol. 14, no. 1, pp. 301–311, 2010.

2. L. Q. Liu and Z. X. Wang, "The development and application practice of wind-solar energy hybrid generation systems in China," Renewable and Sustainable Energy Reviews, vol. 13, no. 6-7, pp. 1504–1512, 2009.

3. Z. S. Li, G. Q. Zhang, D. M. Li, J. Zhou, L. J. Li, and L. X. Li, "Application and development of solar energy in building industry and its prospects in China," Energy Policy, vol. 35, no. 8, pp. 4121–4127, 2007.

4. Z. Jin, W. Yezheng, and Y. Gang, "Estimation of daily diffuse solar radiation in China," Renewable Energy, vol. 29, no. 9, pp. 1537–1548, 2004.

5. Y.-T. Cheng, J.-J. Ho, W. Lee et al., "Efficiency improved by H2 forming gas treatment for Si-based solar cell applications," International Journal of Photoenergy, vol. 2010, Article ID 634162, 6 pages, 2010.

6. M. Taherbaneh, A. H. Rezaie, H. Ghafoorifard, K. Rahimi, and M. B. Menhaj, "Maximizing output power of a solar panel via combination of sun tracking and maximum power point tracking by fuzzy controllers," International Journal of Photoenergy, vol. 2010, Article ID 312580, 13 pages, 2010.

7. H. Li, W. Ma, Y. Lian, and X. Wang, "Estimating daily global solar radiation by day of year in China," Applied Energy, vol. 87, no. 10, pp. 3011–3017, 2010.

8. L. J. Guo, L. Zhao, D. W. Jing et al., "Solar hydrogen production and its development in China," Energy, vol. 34, no. 9, pp. 1073–1090, 2009.

9. J. Zhou, Y. Wu, and G. Yan, "Generation of typical solar radiation year for China," Renewable Energy, vol. 31, no. 12, pp. 1972–1985, 2006.

10. D. H. W. Li and T. N. T. Lam, "Determining the optimum tilt angle and orientation for solar energy collection based on measured solar radiance data," International Journal of Photoenergy, vol. 2007, Article ID 85402, 9 pages, 2007.

11. A. A. Sabziparvar, "General formula for estimation of monthly mean global solar radiation in different climates on the south and north coasts of Iran," International Journal of Photoenergy, vol. 2007, Article ID 94786, 7 pages, 2007.

12. H. Bulut, "Generation of representative solar radiation data for Aegean region of Turkey," International Journal of Physical Sciences, vol. 5, no. 7, pp. 1124–1131, 2010.

13. H. Bulut, O. Büyükalaca, and A. Yilmaz, "Generation of typical solar radiation year for Mediterranean region of Turkey," International Journal of Green Energy, vol. 6, no. 2, pp. 173–183, 2009.

14. Y. Jiang, "Generation of typical meteorological year for different climates of China," Energy, vol. 35, no. 5, pp. 1946–1953, 2010.

15. A. Argiriou, S. Lykoudis, S. Kontoyiannidis et al., "Comparison of methodologies for TMY generation using 20 years data for Athens, Greece," Solar Energy, vol. 66, no. 1, pp. 33–45, 1999.

16. J. Bilbao, A. Miguel, J. A. Franco, and A. Ayuso, "Test reference year generation and evaluation methods in the continental Mediterranean area," Journal of Applied Meteorology, vol. 43, no. 2, pp. 390–400, 2004.

17. A. de Miguel and J. Bilbao, "Test reference year generation from meteorological and simulated solar radiation data," Solar Energy, vol. 78, no. 6, pp. 695–703, 2005.

18. R. Festa and C. F. Ratto, "Proposal of a numerical procedure to select reference years," Solar Energy, vol. 50, no. 1, pp. 9–17, 1993.

19. I. J. Hall, R. R. Prairie, H. E. Anderson, and E. C. Boes, "Generation of a typical meteorological year," in Proceedings of the Annual Meeting of the American Society of the International Solar Energy Society, Calif, USA, June 1978.
20. K. Skeiker, "Comparison of methodologies for TMY generation using 10 years data for Damascus, Syria," Energy Conversion and Management, vol. 48, no. 7, pp. 2090–2102, 2007.
21. S. A. M. Said and H. M. Kadry, "Generation of representative weather-year data for Saudi Arabia," Applied Energy, vol. 48, no. 2, pp. 131–136, 1994.
22. W. Marion and K. Urban, User Manual for TMY2s-Typical Meteorological Years Derived from the 1961–1990 National Solar Radiation Data Base, National Renewable Energy Laboratory,Inc., 1995.
23. M. Petrakis, H. D. Kambezidis, S. Lykoudis et al., "Generation of a "typical meteorological year" for Nicosia, Cyprus," Renewable Energy, vol. 13, no. 3, pp. 381–388, 1998.
24. S. A. Kalogirou, "Generation of typical meteorological year (TMY-2) for Nicosia, Cyprus," Renewable Energy, vol. 28, no. 15, pp. 2317–2334, 2003.
25. T. T. Chow, A. L. S. Chan, K. F. Fong, and Z. Lin, "Some perceptions on typical weather year-from the observations of Hong Kong and Macau," Solar Energy, vol. 80, no. 4, pp. 459–467, 2006.
26. S. Wilcox and W. Marion, Users Manual for TMY3 Data Sets, National Renewable Energy Laboratory,Inc., 2008.
27. L. Yang, J. C. Lam, and J. Liu, "Analysis of typical meteorological years in different climates of China," Energy Conversion and Management, vol. 48, no. 2, pp. 654–668, 2007.
28. J. M. Finkelstein and R. E. Schafer, "Improved goodness-of-fit tests," Biometrika, vol. 58, no. 3, pp. 641–645, 1971.
29. D. Pissimanis, G. Karras, V. Notaridou, and K. Gavra, "The generation of a "typical meteorological year" for the city of Athens," Solar Energy, vol. 40, no. 5, pp. 405–411, 1988.

CHAPTER 2

A Typical Meteorological Year Generation Based on NASA Satellite Imagery (GEOS-I) for Sokoto, Nigeria

OLAYINKA S. OHUNAKIN, MUYIWA S. ADARAMOLA, OLANREWAJU M. OYEWOLA, RICHARD L. FAGBENLE, AND FIDELIS I. ABAM

2.1 INTRODUCTION

Energy remains the convergence point of most critical economic, environmental, and developmental issues confronting the world at the moment. Clean, efficient, stable, and sustainable energy services are ideal for global prosperity. Energy is paramount to achieving Nigeria's Vision 20:2020 needed by the country to be among the top 20 industrialized nations of the world. Lack of energy or its insufficiency in an economy is a potential source of social and economic poverty [1]. In general, a larger proportion of energy is found to be consumed in buildings in Nigeria as is the case in many countries. There is thus a growing concern about energy consump-

A Typical Meteorological Year Generation Based on NASA Satellite Imagery (GEOS-I) for Sokoto, Nigeria. © Ohunakin OS, Adaramola MS, Oyewola OM, Fagbenle RL, and Abam FI. International Journal of Photoenergy *2014 (2014). http://dx.doi.org/10.1155/2014/468562. Licensed under Creative Commons Attribution 3.0 Unported License, http://creativecommons.org/licenses/by/3.0/.*

tion in buildings and the attendant adverse impacts on the environment and with the increasing population and socioeconomic growth (as witnessed in Nigeria), building unit will continue to be a key energy intensive sector [2]. With the country's location on the tropics [3] (latitudes 4°1' and 13°9' North of the Equator and longitudes 2°2' and 14°30' East of the Greenwich Meridian), a large portion of the available energy is consumed on air conditioning. Energy consumption and electric load variation can be very sensitive to weather changes [1]. In Koyuncuolu [4], the unique load variation along with the highly increasing energy consumed for air conditioning called for immediate energy saving programs in which substantial savings can be achieved by decreasing the cooling or heating loads of buildings.

According to Jiang [5], an accurate prediction of building thermal performance relies very much on the weather data such as dry bulb temperature, relative humidity, wind speed, and solar radiation data. Energy simulation in buildings thus offers a valuable tool for engineers and architects to evaluate building energy consumption before the building is constructed; since weather conditions vary significantly from year to year due to the impact of climate change, there is a need to derive a customized weather data set that can well represent the long-term averaged weather conditions over a year which may be referred to as a typical year [4, 5]. Forms of typical years include typical meteorological years (TMY), test reference year (TRY), and the weather year for energy calculations (WYEC). Typical weather year files like TMY are commonly used in building simulations to estimate the annual energy consumptions since analysis using a multiyear dataset is often not feasible and economical for the common design and analysis problems; representative days are too limited and often not accurate enough for a specific design and analysis problem [2]. The typical year approach can thus reduce the computational efforts in simulation and weather data handling by using one year instead of multiple years. However, in Skeiker and Abdul Ghani [6], the term "typicality" could be interpreted in many ways as used by several investigators, namely, (i) the selection of weather which appears to be typical of an appropriate portion of the year, (ii) selection of a year which appears to be typical from several years of solar radiation data, and (iii) some investigators having run long periods of observational data in an attempt to simulate typical weather for the calculation.

Several works have been carried out on the development of TMY for various locations worldwide (e.g., [7–14]) whereas in Nigeria, existing work on the generation of weather data may be limited to (i) TRY generation for Ibadan, (ii) TMY generation for Port Harcourt zone, and (iii) generation of a typical meteorological year for northeast, Nigeria [1, 15, 16]. The primary aim of this work is to generate TMY for Sokoto (located in North Nigeria with latitude—13.01°N, longitude—05.15°E, and altitude—350.8 m) using a 23-year (1984–2006) hourly weather data from available satellite data acquired from NASA Geostationary Operational Environmental Satellite-1 (GEOS-1) *Multiyear Time-Series Data*. Figure 1 shows the studied site (Sokoto) and the other locations for which TMYs have been previously generated. Satellite and model-based products have been shown to be accurate enough to provide reliable solar and meteorological resource data over regions where surface measurements are sparse or nonexistent [17, 18]. This advantage of wide GEOS-1 data coverage has been utilized in the work of Dike et al., [19] for correlating meteorological estimates for seventeen locations within the southern region of Nigeria while Fadare [20] adopted the same dataset for modeling solar energy potential in Nigeria.

The selected weather data (including global solar radiation, dew point temperature, mean temperature, maximum temperature, minimum temperature, relative humidity, and wind speed) were chosen due to their impact on (and significant contribution to) the solar energy systems performance and heat gain and loss in buildings. For instance, in building applications, the knowledge of solar radiation is vital for accurate determination of cooling load in tropical regions (like Nigeria). It should also be noted that the air ambient temperature influences the thermal response of a building and the amount of heat gain and loss through it. The relative humidity is essential for the determination of latent heat for air-conditioning systems and evaporation levels. For solar energy conversion systems (Concentrated Solar Power and Photovoltaic systems), these meteorological data are also essential for design, selection, and performance evaluation of these systems.

FIGURE 1: Map of Nigeria showing the studied site and locations with existing TMY values.

2.2 MATERIALS AND METHOD

A TMY consists of 12 typical meteorological months (TMMs) selected from various calendar months in a multiyear weather database [5]. Several methods including Festa and Ratto, Danish and Sandia commonly exist for the development of TMY [21–23]. Sandia method is adopted in this work because of its wide applicability in the generation of weather data in tropical countries when compared with other methods [24]. However, Sandia procedure is difficult to implement directly in spite of its wide deployment in generating TMY [25]. In view of this, several procedures were described in literatures for generating TMY using Sandia method [13–15, 25]; this work employed the procedural steps developed and outlined in Sawaqed et al., [25] due to its simplicity, for the development of TMY for Sokoto.

In Sawaqed et al., [25], TMY development begins with the generation of typical meteorological month (TMM). The TMM procedure involves the selection of five candidate years that are closest to the composite of all the years under study for each of the 12 calendar months. The statistical analysis commenced with the generation of daily weather indices from the adopter parameters for each day of record. Each of the seven sets of daily indices is sorted into bins by month and they are used to establish 12 long-term cumulative distribution functions (CDFs) given by [25]

$$CDF_j = \frac{1}{n}j, \quad j = 1,2,\dots n \tag{1}$$

where j is ranked order number and n is the total number of elements. A nonparametric method known as the Finkelstein-Schafer (FS) statistics [26] is then used to measure the closeness between the corresponding short-term and long-term CDFs as using (2) [26]:

$$FS_j = \frac{1}{n}\sum_{i=1}^{n}\delta_i, \quad i = 1,2,\dots,n \tag{2}$$

where δ_i is the absolute difference between the short-term and the long-term CDFs for day i in the month, n is the number of days in the month, and j is the parameter (index) considered. For mean temperature j = 1, for max temperature, j = 2, and for solar flux j = 7. For each of the candidate months, the seven different FS statistics calculated for the seven indices are grouped into a composite weighted index using (3) [25] together with the weighting factors listed in Table 1. The resulting product gives the weighted sum (WS):

$$WS = \sum_{j=1}^{m} w_j FS_j \tag{3}$$

where m is the number of indices (parameters/elements) considered, W_j is the weight for index j, FS_j is the FS statistic for index j.

TABLE 1: Weighting factors of FS statistics.

Parameter	Weight (w_j)
DBT (minimum)	1/12
DBT (maximum)	1/12
DBT (mean)	2/12
Dew point (mean)	1/12
Relative humidity	1/12
Solar radiation	6/12

A weighting factor of 0.5 is assigned to global solar radiation while other weather indices were given equal weights that add up to 0.5. This is because (i) in tropical region with warm weather (like Nigeria), solar heat gain can be more significant in cooling load calculations for building as well as for passive and active heating systems and (ii) other weather parameters are directly or indirectly affected by global solar radiation.

All individual months are ranked in ascending order of the WS values. A typical month is then selected by choosing from among the five months with the lowest WS values by following the procedure developed by Sawaqed et al. [25]. The persistence of mean dry bulb temperature and daily global horizontal solar radiation are evaluated by determining the frequency and run length above and below fixed long-term percentiles. For mean daily dry bulb temperature, the frequency and run length above the 67th percentile and below the 33rd percentile are determined. For global horizontal radiation, the frequency and run length below the 33rd percentile are also determined. The persistence data are used to select, from the five candidate months, the month to be used in the TMY. The highest ranked candidate month in ascending order of the WS values that meet the persistence criterion is used in the TMY.

The generated TMY values were compared with the long-term average values for the location under study. The performance of the TMY was investigated using methods of stochastic analysis to calculate the root-mean-square error (RMSE) given in (4) as expressed in Ohunakin et al. [27]:

$$RMSE = \left[\frac{\Sigma\left(H_{pred} - H_{obs}\right)^2}{n} \right]^{1/2}$$

(4)

2.3 RESULTS AND DISCUSSION

In this work, the mean temperature and global solar radiation are used in selecting 12 typical meteorological months (TMMs) and together with the remaining parameters, TMY for Sokoto is developed. For each month, the FS statistic is also estimated for every year and for all of the six parameters that have been considered. For example, the FS statistics of global solar radiation for Sokoto is presented in Table 2. It can be observed from this table that the FS statistics show variation from one month to the other. Similar variation is also observed among other variables.

TABLE 2: FS statistics for global solar radiation.

	Jan	Feb	Mar	Apr	May	Jun	Jul	Aug	Sept	Oct	Nov	Dec
1984	2.621	1.289	1.379	2.200	1.736	2.305	3.708	4.748	3.355	2.618	2.296	0.929
1985	0.844	1.046	1.663	2.055	1.532	2.381	4.091	4.835	4.180	2.766	2.497	1.586
1986	2.410	1.907	1.535	2.316	2.443	1.704	4.826	4.033	3.883	3.230	2.086	0.949
1987	0.678	1.019	2.203	2.779	2.705	2.085	4.130	3.695	4.243	2.422	2.989	0.732
1988	1.577	2.090	2.134	3.311	1.947	2.235	3.820	4.319	3.538	2.223	3.212	0.826
1989	2.203	0.853	2.121	2.623	2.874	2.381	4.073	2.850	2.311	1.720	1.657	2.107
1990	1.073	1.572	1.972	2.250	2.861	2.045	4.466	3.090	2.456	1.954	1.903	1.091
1991	0.659	0.765	2.681	3.169	4.683	1.643	3.829	3.671	2.724	1.872	2.054	1.293
1992	1.094	0.733	2.861	2.103	2.490	2.020	4.737	2.116	2.657	1.646	2.747	1.387
1993	0.729	1.462	1.407	1.795	1.716	2.052	3.239	3.677	1.989	1.367	1.906	0.945
1994	2.249	0.788	1.192	1.703	2.463	2.425	3.850	3.163	2.353	2.702	1.904	0.833
1995	1.027	0.575	1.251	3.526	3.232	2.718	3.755	2.750	2.487	1.376	1.676	0.866
1996	0.801	0.637	1.104	2.686	2.216	2.350	3.512	3.096	2.592	1.668	1.728	0.564
1997	0.409	0.505	2.800	1.823	3.413	2.711	4.527	3.695	1.693	1.643	2.134	1.109
1998	0.450	0.858	0.804	2.152	1.933	2.092	4.849	3.718	2.733	1.743	1.860	0.936
1999	0.918	2.630	1.504	2.473	2.677	2.893	3.780	3.915	2.843	2.176	1.978	0.894
2000	1.124	1.396	0.759	1.933	1.724	3.024	3.302	2.637	3.456	2.711	1.382	0.984
2001	0.834	0.829	1.087	1.681	2.679	2.838	3.554	3.012	2.818	1.129	1.870	0.660
2002	0.356	0.636	1.481	1.941	1.863	2.267	4.226	3.336	2.229	2.276	1.774	0.675
2003	0.616	1.383	1.352	3.188	3.003	3.986	3.868	4.642	2.554	2.070	1.780	0.922
2004	1.106	1.334	2.139	2.314	2.583	2.694	4.385	2.942	2.170	1.742	1.859	1.479
2005	1.479	0.787	2.069	2.784	3.490	3.808	3.833	3.138	1.976	1.479	1.874	0.890
2006	0.815	0.875	1.118	2.386	3.039	2.865	3.083	3.107	3.528	2.176	3.511	1.541

TABLE 3: WS of FS statistics for the indices.

	Jan	Feb	Mar	Apr	May	Jun	Jul	Aug	Sept	Oct	Nov	Dec
1984	1.770	1.383	1.958	3.499	4.219	2.908	2.641	2.853	2.661	4.273	2.147	1.026
1985	1.497	1.636	3.620	2.924	4.466	3.087	2.896	2.870	2.753	2.645	1.701	1.332
1986	1.739	1.874	3.180	3.665	4.167	2.606	3.370	2.559	2.920	3.529	2.725	1.198
1987	0.984	1.155	3.121	3.891	4.124	2.520	2.609	2.461	2.845	3.400	1.823	1.075
1988	1.666	1.656	2.696	3.842	3.500	2.861	2.799	2.959	2.816	2.569	2.128	1.350
1989	2.222	1.387	2.392	2.617	4.599	2.585	2.493	1.871	1.457	2.533	1.368	1.838
1990	1.606	1.488	2.669	2.918	3.928	2.427	2.934	1.969	1.625	2.640	1.559	2.274
1991	0.965	2.406	2.339	3.160	5.363	1.978	2.662	2.321	2.487	1.896	1.534	1.201
1992	1.834	1.234	2.468	2.521	3.855	2.561	2.094	1.612	2.472	2.968	2.101	1.313
1993	1.448	1.356	2.195	2.711	2.953	2.620	2.545	2.516	1.579	2.388	1.685	1.125
1994	2.143	1.286	1.683	1.901	3.316	4.386	2.914	1.952	1.773	1.919	1.812	1.265
1995	1.541	0.971	1.575	3.137	3.987	2.716	2.594	1.817	1.635	1.460	1.657	1.483
1996	1.091	1.179	1.520	2.855	3.068	2.102	2.234	2.014	1.491	2.036	1.756	0.989
1997	0.854	1.482	2.996	2.538	3.238	2.584	2.995	2.465	1.588	1.670	1.856	1.271
1998	1.331	1.129	2.139	2.732	2.615	2.464	3.165	2.417	2.351	3.473	1.477	1.365
1999	1.150	2.538	2.412	2.272	2.822	3.843	2.467	2.766	2.106	2.173	1.416	1.120
2000	1.748	1.746	1.898	1.746	2.581	3.245	3.109	2.124	2.337	2.383	1.476	1.276
2001	1.364	1.326	1.939	2.719	2.929	2.610	2.483	1.978	1.953	2.242	1.650	1.048
2002	0.896	1.267	1.918	1.944	2.634	3.756	3.024	2.019	1.613	2.893	2.116	1.245
2003	1.326	2.269	2.196	3.378	4.278	3.043	2.414	2.911	2.131	1.989	1.860	1.273
2004	1.456	1.640	3.125	2.288	3.106	3.559	4.508	3.703	3.197	3.095	1.851	1.498
2005	1.998	2.289	2.962	2.570	3.477	3.474	2.697	2.929	2.655	2.338	1.475	1.260
2006	1.840	1.782	2.088	3.455	3.227	2.842	2.034	2.189	2.708	4.470	2.413	1.436

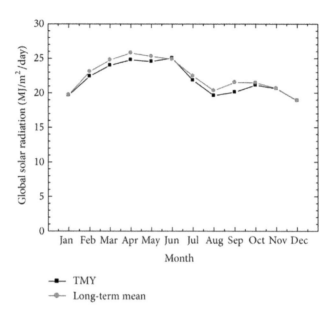

FIGURE 2: Predicted TMY with long-term average values for the global solar radiation.

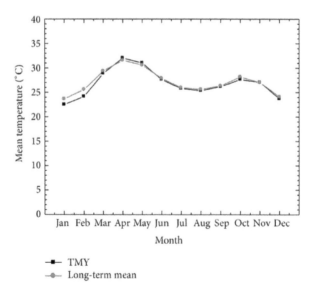

FIGURE 3: Predicted TMY with long-term average values for the mean temperature.

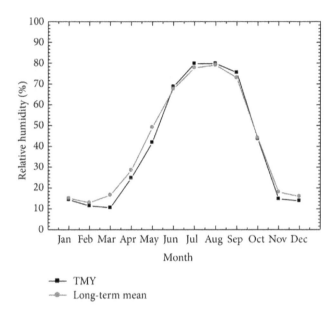

FIGURE 4: Predicted TMY with long-term average values for the relative humidity.

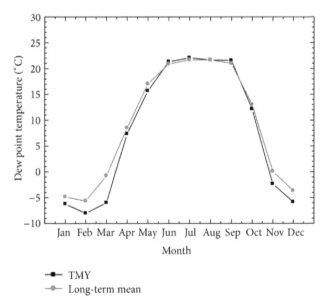

FIGURE 5: Predicted TMY with long-term average values for the dew point temperature.

With the daily values of the selected parameters determined from the measured data, the values of the WS for each month of the 23-year period were then calculated as shown in Table 3. The five candidate years for each months having the lowest values of WS together with the corresponding year are all arranged in ascending order. The persistence of mean dry bulb temperature and daily global horizontal radiation on the five candidate months were evaluated by determining the frequency and run length above and below fixed percentiles (33rd and 67th) as discussed in [8]. The selected 12 TMMs are shown in Table 4. It can be seen that the 12 TMMs spread across the whole 23-year period; three TMMs are found in 1996 and two in 1999, whereas only one TMM exists in 1989, 1992, 1994, 1995, 2000, 2001, and 2002, respectively.

TABLE 4: Selected typical meteorological months.

Month	Year
January	2002
February	1992
March	2001
April	1999
May	1999
June	1996
July	1989
August	1994
September	1995
October	1996
November	2000
December	1996

The relationship between the long-term mean and the TMY for selected weather parameters are presented in Figures 2–5.

Figure 2 showed that good relationships exist between the long-term mean (23 year) and the TMY values for the global solar radiation, though with very small variations. It can be observed that the TMY and the long-term solar fluxes exceed $24\,MJ/m^2/day$ from March to June with the peak as 25.1 and 25.8 MJ/m^2/day in June and April for the TMY and long-term fluxes, respectively. Figure 3 shows the monthly average mean temperature based on TMY and long-term values. It can be seen that Figure 3 follows similar trend with Figure 2. The peak values of the mean temperature are observed in April as 32.0 and 31.6°C for the TMY and long-term average values, respectively. The good agreement between TMY and long-term global solar radiation and mean temperature may be related to high weighting factor placed on each parameter being 0.5 for global solar radiation and 2/12 for mean temperatures (Table 1). The root-mean-square error (RMSE), which measures level accuracy between the TMY and long-term, is estimated as 0.62°C and 0.68 MJ/m^2/day for the mean temperature and global solar radiation. These low values further confirm good agreement between the TMY and long-term values for these parameters.

The relationship between the relative humidity (Figure 4) and dew point temperature (Figure 5) is reasonable but not as close as that for the global solar radiation and the mean temperature. The RMSE values for the relative humidity and dew point temperature are 3.23% and 2.04°C, respectively. These values are relatively small, but higher than RMSE values for the mean temperature and global solar radiation. This may be due to nonuniformity of the weighting factors and may also result from the fact that the selection of the TMMs is based on persistence of mean temperature and daily global horizontal radiation. Hence, the TMMs selected for the TMY can represent the long-term meteorological weather conditions for the Sokoto as considered in this study.

2.4 CONCLUSIONS

The TMY generated will be very useful for optimal design and evaluation of solar energy conversion systems, HVAC, and other solar energy dependent systems to be located within the vicinity of Sokoto. This work is a part of ongoing research work to develop a comprehensive TMY weather database for hourly building energy simulations in Nigeria and design of solar energy systems. In addition, the adopted satellite data have significantly advanced our ability to observe weather systems by providing frequent interval visible and infrared imagery of the earth surface, atmospheric moisture, and cloud cover that otherwise would have been impossible with ground observation. Findings based on the adoption of Filkenstein-Schafer statistical method utilized for the creation of a TMY for Sokoto show the following.

1. The developed 12 TMMs are evenly spread across the whole 23-year period with three TMMs found in 1996 and two in 1999, and in 1989, 1992, 1994, 1995, 2000, 2001, and 2002 only one exists.
2. The close fit observed in the values suggests a close agreement between the TMY predictions and long-term averages.
3. Observed agreement between TMY and long-term values of the respective fluxes may be related to weighting factor assigned to each parameter.

REFERENCES

4. O. S. Ohunakin, M. S. Adaramola, O. M. Oyewola, and R. O. Fagbenle, "Generation of a typical meteorological year for north-east, Nigeria," Applied Energy, vol. 112, pp. 152–159, 2013.
5. L. Yang, J. C. Lam, and J. Liu, "Analysis of typical meteorological years in different climates of China," Energy Conversion and Management, vol. 48, no. 2, pp. 654–668, 2007.
6. O. S. Ohunakin, M. S. Adaramola, O. M. Oyewola, and R. O. Fagbenle, "Solar energy applications and development in Nigeria: drivers and barriers," Renewable and Sustainable Energy Reviews, vol. 32, pp. 294–301, 2014.
7. H. Koyuncuolu, "Dynamic energy analysis in buildings," http://www.fbe.deu.edu.tr/ALL_FILES/Tez_Arsivi/2004/YL_p1724.pdf.
8. Y. Jiang, "Generation of typical meteorological year for different climates of China," Energy, vol. 35, no. 5, pp. 1946–1953, 2010.
9. K. Skeiker and B. A. Ghani, "Advanced software tool for the creation of a typical meteorological year," Energy Conversion and Management, vol. 49, no. 10, pp. 2581–2587, 2008.
10. D. Pissimanis, G. Karras, V. Notaridou, and K. Gavra, "The generation of a "typical meteorological year" for the city of Athens," Solar Energy, vol. 40, no. 5, pp. 405–411, 1988.
11. A. Argiriou, S. Lykoudis, S. Kontoyiannidis et al., "Comparison of methodologies for TMY generation using 20 years data for Athens, Greece," Solar Energy, vol. 66, no. 1, pp. 33–45, 1999.

12. S. A. M. Said and H. M. Kadry, "Generation of representative weather—year data for Saudi Arabia," Applied Energy, vol. 48, no. 2, pp. 131–136, 1994.

13. J. C. Lam, S. C. M. Hui, and A. L. S. Chan, "A statistical approach to the development of a typical meteorological year for Hong Kong," Architectural Science Review, vol. 39, no. 4, pp. 201–209, 1996.

14. Q. Zhang, J. Huang, and S. Lang, "Development of the typical and design weather data for Asian locations," Journal of Asian Architectural Building Engineering, vol. 1, no. 2, pp. 49–55, 2002.

15. D. J. Thevenard and A. P. Brunger, "The development of typical weather years for international locations: part I, algorithms," ASHRAE Transaction, vol. 108, no. 2, pp. 376–383, 2002.

16. D. J. Thevenard and A. P. Brunger, "The development of typical weather years for international locations: part II, production," ASHRAE Transaction, vol. 108, no. 2, pp. 480–486, 2002.

17. Q. Zhang, J. Huang, and S. Lang, "Development of typical year weather data for Chinese locations," ASHRAE Transaction, vol. 108, no. 2, pp. 1063–1075, 2002.

18. R. Layi Fagbenle, "Generation of a test reference year for Ibadan, Nigeria," Energy Conversion and Management, vol. 36, no. 1, pp. 61–63, 1995.

19. C. O. C. Oko and O. B. Ogoloma, "Generation of a typical meteorological year for port harcourt zone," Journal of Engineering Science and Technology, vol. 6, no. 2, pp. 204–221, 2011.

20. SSE6 Methodology, "Surface meteorology and solar energy, (SSE) release 6.0 methodology," https://eosweb.larc.nasa.gov/cgi-bin/sse/sse.cgi?+s01+s05+s06#s05.

21. O. S. Ohunakin, Fluid dynamics modelling of the impact of climate change on solar radiation in Nigeria [Ph.D. thesis], 2013.

22. V. N. Dike, T. C. Chineke, O. K. Nwofor, and U. K. Okoro, "Evaluation of horizontal surface solar radiation levels in southern Nigeria," Journal of Renewable and Sustainable Energy, vol. 3, no. 2, Article ID 023101, 2011.

23. D. A. Fadare, "Modelling of solar energy potential in Nigeria using an artificial neural network model," Applied Energy, vol. 86, no. 9, pp. 1410–1422, 2009.

24. I. J. Hall, R. R. Prairie, H. E. Anderson, and E. C. Boes, "Generation of typical meteorological year for 26 SOLMET stations," Sandia Laboratories Report SAND 78-1601, Sandia Laboratories, Albuquerque, NM, USA, 1978.

25. H. Lund and S. Eidorff, Selection Methods for Production of Test Reference Years, Appendix D, Contract 284-77 ES DK, Report EUR, Commission of the European Communities, 1980.

26. R. Festa and C. F. Ratto, "Proposal of a numerical procedure to select reference years," Solar Energy, vol. 50, no. 1, pp. 9–17, 1993.

27. S. Janjai and P. Deeyai, "Comparison of methods for generating typical meteorological year using meteorological data from a tropical environment," Applied Energy, vol. 86, no. 4, pp. 528–537, 2009.

28. N. M. Sawaqed, Y. H. Zurigat, and H. Al-Hinai, "A step-by-step application of Sandia method in developing typical meteorological years for different locations in Oman," International Journal of Energy Research, vol. 29, no. 8, pp. 723–737, 2005.

29. J. M. Filkenstein and R. E. Schafer, "Improved goodness-of-fit tests," Biometrika, vol. 58, no. 3, pp. 641–645, 1971.

30. O. S. Ohunakin, M. S. Adaramola, O. M. Oyewola, and R. O. Fagbenle, "Correlations for estimating solar radiation using sunshine hours and temperature measurement in Osogbo, Osun State, Nigeria," Frontiers in Energy, vol. 7, no. 2, pp. 214–222, 2013.

PART II

WIND ENERGY AND THE WEATHER

CHAPTER 3

Assessing Wind Farm Reliability Using Weather Dependent Failure Rates

G. WILSON AND D. MCMILLAN

3.1 INTRODUCTION

A wind turbine's reliability is commonly assessed based on the availability it achieves. The availability is the proportion of time in which an asset is able to produce electricity [1]. The time in which a turbine is unavailable is due to either corrective or preventative maintenance [2][3]. Onshore wind turbines typically perform very well, achieving availabilities between 97 %—99 % [4].

To increase the availability of a wind turbine more money can be spent on operation and maintenance (O&M) with the aim of reducing the probability of a significant downtime occurring. However the cost of O&M must be weighed up against the cost of lost revenue, which is due to downtime [5]. By increasing the cost of O&M there is a reduction in downtime,

Assessing Wind Farm Reliability Using Weather Dependent Failure Rates . © Wilson G and McMillan D. Journal of Physics: Conference Series *524 (2014). doi:10.1088/1742-6596/524/1/012181. Licensed under a Creative Commons Attribution 3.0 Unported License, http://creativecommons.org/licenses/ by/3.0/.*

but a point is reached where the direct cost of O&M is greater than the savings made by increasing the availability of the asset. This is the point when it no longer makes financial sense to increase O&M spending.

It has been demonstrated that there is a relationship between wind speed and wind turbine failure rates [6]–[8]. Previous research has demonstrated that wind turbine component failure rates can be calculated as a function of wind speed [9][10]. Using this methodology it is possible to model how failures to wind turbine components, and the resultant downtimes, effect energy production and the cost of O&M.

This paper aims to assess the reliability and productivity of three potential wind farm sites using wind speed dependent failure rates—for which a method has been developed in previous research [9][10]. A comparison will then be made between the expected output according to a model that incorporates the wind speed dependent failure rates and one which does not.

TABLE 1: Component failure rate and data points

Components	Failure Rate (Failures per wind turbine year)	Failure log entries/data points
Emergency Systems	0.0260	2
Met Instruments	0.0754	29
Rotor	0.0468	18
Blade Pitch System	0.0676	26
Drive Train	0.1561	60
Yaw System	0.1509	58
Hydraulic System	0.0780	30
Control System	0.5203	200
Main Generator	0.0312	12
Lifting System	0.0104	4
Nacelle	0.0156	6
Tower	0.0598	23
Total	1.2381	468

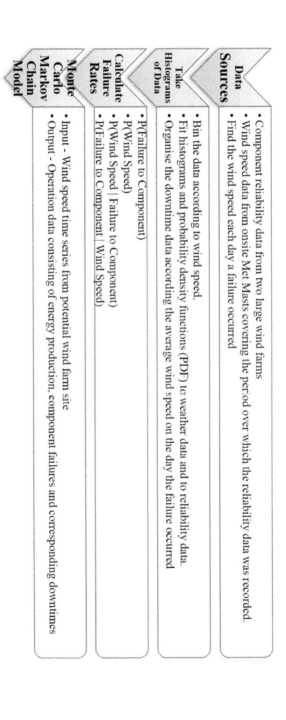

FIGURE 1: Methodology

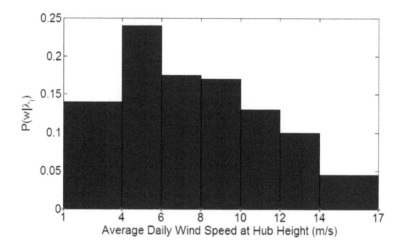

FIGURE 2: The probability of a wind speed w occurring on a given day, when a failure has occurred to i.

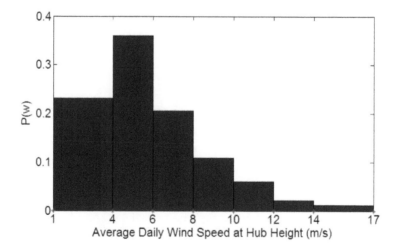

FIGURE 3: The probability of a wind speed w occurring on a given day.

3.2 METHODOLOGY

The methodology in this paper is outlined in Figure 1. The datasets come from a reliability record of two onshore wind farms (sites 1 and 2), which comprised of 468 individual failures throughout 381.7 wind turbine years of operation, and meteorological masts which measured the wind speed on each of the sites throughout the same period. The wind turbines on both sites are the same model and are between 3 and 6 years old. They are variable speed, pitch regulated and have a generation capacity of 2.3MW. Only accounting for downtime attributed to corrective maintenance, the average availability of the two sites over the recorded period was 99.44 %. Assuming two days of preventative maintenance takes place per wind turbine year this reduces to 98.89 % availability.

The average daily wind speeds of sites 1 and 2 are 5.86m/s and 6.62m/s respectively [10]. Data comes from two onsite met masts—for both Blacklaw and Whitelee—that are assumed to take measurements that are representative of hub height wind speed across the whole of both sites. The wind speed data and reliability data from both sites were combined to produce a dataset which matched each recorded failure with the wind speed that occurred on that site on the day of failure. This dataset was then split up into the twelve main components, shown in Table 1 with their respective failure rates calculated using the reliability data. The components which failed the most frequently are the control system, the yaw system and the drive train.

3.2.1 CALCULATING THE FAILURE RATES

To calculate a component failure rate as a function of wind speed, Bayes Theorem is used; this is shown in equation 1.

$$P(\lambda_i|w) = \frac{P(w|\lambda_i)P(\lambda_i)}{P(w)} \tag{1}$$

If w is the average daily wind speed on a given day, λ_i represents a failure to component i. The term on the left hand side of equation 1 represents the probability of a failure occurring to a component (λ_i), given the average daily wind speed w. This probability represents the failure rate of component i, as a function of average daily wind speed w.

The terms on the right hand side of equation 1 can be calculated from the reliability data and the met mast data. The term $P(w|\lambda_i)$, is the probability of wind speed w occurring, given a failure has occurred to component i. This is calculated by taking a normalised histogram (PDF) of the daily average wind speeds recorded on days when a failure occurred to component i. The data is binned as shown in Figure 2 and Figure 3. The highest daily average wind speed measured from sites 1 and 2 was 17 m/s.

$$\lambda_i = \frac{number\ of\ failures\ in\ given\ period\ to\ component\ i}{total\ operation\ time\ (days)} \qquad (2)$$

The second numerator term is (λ_i), which is the probability of a failure occurring to component i on a given day, otherwise known as the daily failure rate of i. This is calculated by using equation 2. The annual failure rates for each component are shown in Table 1.

The denominator term P(w) represents the probability that the daily average wind speed is w. Assuming that daily average wind speeds have the same effect on the failure rates of components on both sites, this is calculated by taking a second PDF, using the same bins as before, of the daily average wind speeds recorded on sites 1 and 2 throughout the period that the reliability data was recorded. The quantity of data from site 1 and 2 differ, so a merged data set was produced which describes the average daily wind speed of both sites proportionally, this is shown in Figure 3. A more detailed explanation of this merging can be found in Wilson and McMillan [10]. The average wind speed of this merged dataset is 5.98 m/s.

Therefore with P(w), P(λ_i) and P(w|λ_i) known, equation 1 was used to calculate the probability of a failure to component i, given a daily average wind speed w, P(λ_i|w). The wind speed dependent failure rates for the control system, yaw system and drive train are shown in Figure 4.

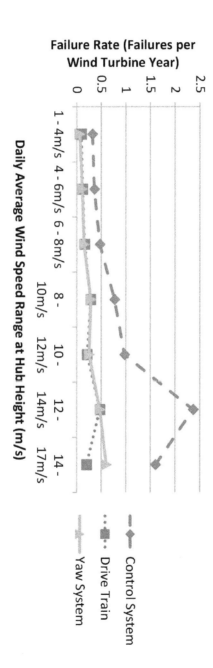

FIGURE 4: Wind speed dependent failure rates

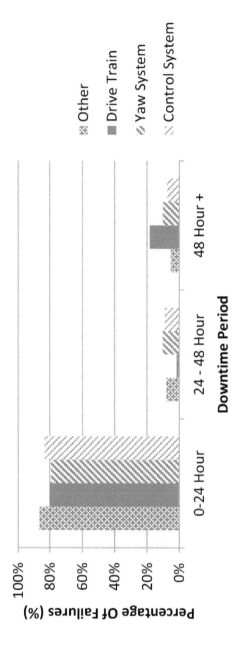

FIGURE 5: Distribution of downtime

3.2.2 DOWNTIME

Every failure recorded in the reliability record has a downtime attributed to it. The downtimes were organised similarly to the recorded failures, first they were split up according to the failed component and then were divided up into the wind speed bins, according to the average daily wind speed when the failure occurred. Figure 5 shows the distribution of downtimes for the control system, yaw system, drive train and the remaining nine components.

Of the failures in the dataset, 83.3% had a downtime of less than 24 hours; this is an improvement on the downtime distribution calculated by Faulstich et al [11]. This was to be expected as this dataset comprises of a larger and more modern and advanced wind turbine model.

3.2.3 MONTE CARLO MARKOV CHAIN SIMULATION

Often the reliability of engineering systems is described as being discrete in that it can exist in one state until a transition occurs and the system changes to another state. These transitions can be represented by a transitional probability matrix. This characteristic means that many systems can be modelled as a Markov process. Markov Chains have been used often to model components and systems [12][13][14][15][16].

Each component in this analysis is represented by a markov chain that can exist in one of two states at any time, "operating" or "failed". The transition rate between the "operating" state and the "failed" state is the failure rate (λ). If a failure occurs to any component, the whole system stops operating and suffers a downtime before then returning to the "operating" state.

A Markov Chain Monte Carlo (MCMC) model has been developed which—using the wind speed dependent failure rates calculated in section 2.1, as opposed to the static failure rate in Table 1—determines the impact of wind speeds on the availability of the wind turbine and its components throughout its lifetime. MCMC has been used previously by various authors to model reliability [2][11][14][15]. The lifespan of a wind farm is then simulated using historical wind speed data from a potential wind farm site as a model input. From this simulated operation data, component failure rates, wind farm reliability and energy production can be evaluated.

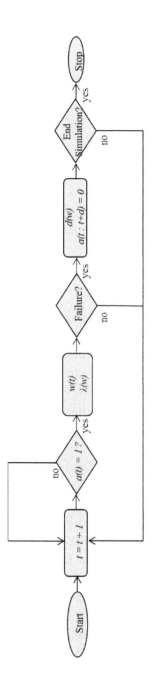

FIGURE 6: Markov Chain Monte Carlo simulation method.

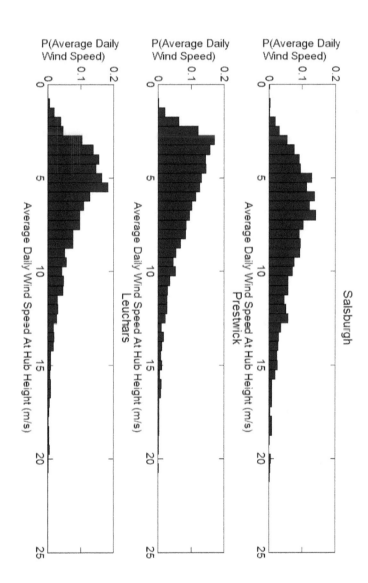

FIGURE 7: Wind speed distributions for Salsburgh, Prestwick and Leuchars.

In the MCMC simulation, the downtime a component spends in the "failed" state before it returns to the "operating" state is calculated by sampling according to which wind speed bin the average daily speed on the day of failure fits into. If for example the control system fails when the daily wind speed falls between 8 m/s and 10 m/s, there are 34 recorded control system failures which have occurred in the data in that wind speed bin, all with corresponding downtimes. A downtime is selected from this set randomly with uniform probability distribution using Monte Carlo Simulation.

Figure 6 summarises the simulation method where t represents time in days since the start of the simulation, a denotes the availability of the wind turbine, which is either available (a = 1) or unavailable (a = 0) at time t. The downtime and failure rate are, respectively, d and λ. Both d and λ are functions of the wind speed, w which is a function of time, t. The simulation continues running for 200,000 wind turbine years until the variance of the samples become constant.

3.3 RESULTS AND DISCUSSION

The average wind speeds of the potential wind farm sites, Salsburgh, Prestwick and Leuchars are 7.91 m/s, 6.16 m/s and 6.59 m/s respectively. The data for each site comes from the Met Office [17] and consists of three years of hourly wind speed data—between 2009 and the end of 2011—recorded using 10m masts, which has been aggregated into daily average wind speed. The wind speed profile power law is used to extrapolate the 10m wind speeds to a hub height of 82m, using a surface roughness factor of 0.04 [18]. The wind speed distributions of the sites are shown in Figure 7.

3.3.1 RELIABILITY

Using the model, the failure rate of the three potential wind farms throughout the year is calculated. The mean results from the MCMC simulation is shown in Figure 8. As the three potential sites are all located in Scotland, they followed approximately the same seasonal trend in that the failure rate declined in the summer periods when the wind was calmer and produced higher failure rates in the winter when the wind speed was high. February had lower wind speeds in Scotland than usual during 2009–2011 which explains the low failure rate for all three sites in February.

TABLE 2: Calculated site availability and system failure rate

Site	Wind Farm Failure Rate (Failures per wind turbine year)	Availability (%)
Salsburgh	1.60	98.65
Prestwick	1.29	98.84
Leuchars	1.34	98.81

Table 2 shows the annual wind turbine failure rates and availabilities calculated by the model, assuming that the wind turbines on each wind farm are shut down two days a year for preventative maintenance. The site with the highest availability is Prestwick, which also had the lowest average daily wind speed. Equally the site which had the highest wind speed, Salsburgh, has the highest failure rate and lowest availability.

The individual components all react differently to wind speed as shown in Figure 4. The effect of the wind speed on the individual components is shown in Figure 9. The components which are most badly affected by increased wind speeds are the control system and yaw system. The site which differs most to the reliability data is Salsburgh—with the failure rates of the control system and yaw system increasing by 44% and 45% respectively. The drive train failure rate increases by 26%, 2% and 6% on Salsburgh, Prestwick and Leuchars respectively.

3.3.2 ECONOMIC ASSESSMENT

The aim of this section is to assess the economic benefit of using the wind speed dependent failure rate MCMC model as opposed to simply assuming static failure rates and availabilities which are calculated from the reliability data of and are not wind speed dependent.

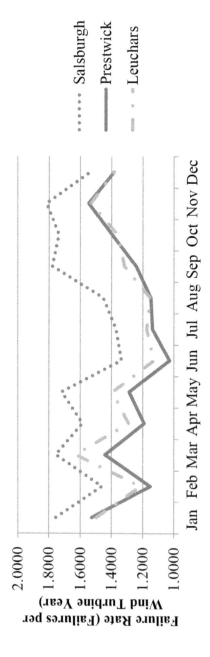

FIGURE 8: Site seasonal failure rates.

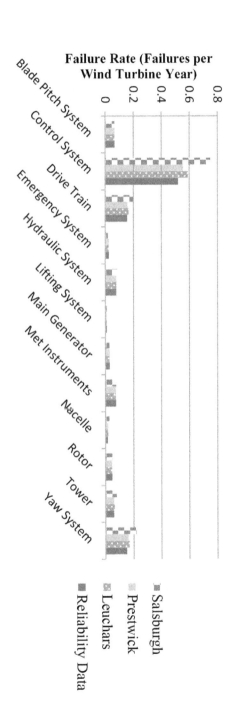

FIGURE 9: Component failure rates for the three sites calculated by the model compared to the original reliability data.

TABLE 3: Variables used in calculation of net profit

Availability	a	98.89%
O&M Cost per Turbine Year	β	£31,242
Unit Price of Electricity	s	80 £/MWh
Capital Expenditure	CapEx	£3,000,000
Lifetime of Site	l	7300 days

Using this static model, the net profit of a site can be calculated using equations 3-5 and the taking the values in Table 3. This static method assumes therefore that each site (using the same model of turbines as site 1 and 2) will be equally reliable and will therefore have the same component failure rates, site availability and O&M cost, regardless of the wind speed distribution of the site.

The availability and O&M cost in Table 3 have been calculated from the reliability data and are used in equations 3 and 4. The unit price of electricity in Table 3 is broken down to £40/MWh as a conservative estimate of the price of a unit of electricity sold on the UK market and the UK subsidy (the Renewable Obligation Certificate) which is also estimated to be roughly £40/MWh for onshore wind [19][20]. The power curve shown in Figure 10 is used to calculate generated power P(U(t)) and is used by both the wind speed dependent MCMC model and the static model.

$$Revenue = a \sum_{t=1}^{l} 24P\big(U(t)\big)s \tag{3}$$

$$OpEx = \frac{\beta l}{365} \tag{4}$$

$$Net = Revenue - CapEx - OpEx \tag{5}$$

Unfortunately replacement part costs and labour costs are not available for the reliability dataset. O&M costs are typically estimated to be 0.6 - 0.7 c€/kWh [21]. Based on the production of sites 1 and 2 this equates to approximately £30,000 - £35,000 per wind turbine year.

To calculate approximate costs of O&M, which take into account wind turbine reliability and not production, downtime is considered as it is known to correlate with failure severity and therefore cost [22]. The downtimes are grouped as 0–24 hours, 24–48 hours and over 48 hours; these are assumed to cost £2,500, £25,000 and £250,000 respectively per failure, not including lost production. Applying these assumed costs to the reliability data results in an O&M cost of £31,242 per wind turbine year. This is the cost of O&M in the static model and compares well with estimates made in Morthorst [21].

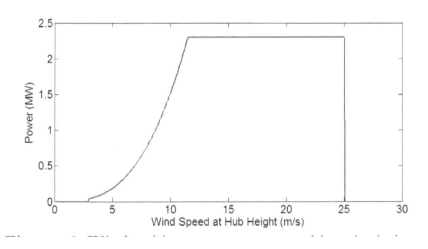

FIGURE 10: Wind turbine power curve used in calculations.

TABLE 4: Comparison between model and equation 5

Site	Revenue (per turbine)			O&M (per turbine Year)			Net (per turbine)		
	MCMC (£millions)	Static -Equation 3 (£millions)	Difference (MCMC - Static)	MCMC (£)	Static - β (£)	Difference (MCMC - Static)	MCMC (£millions)	Static - Equation 5 (£millions)	Difference (MCMC - Static)
Salsburgh	12.43	12.52	-0.72%	49313	31242	57.84%	8.44	8.90	-5.07%
Prestwick	7.66	7.67	-0.13%	34787	31242	11.35%	3.96	4.05	-2.00%
Leuchars	8.58	8.59	-0.12%	36742	31242	17.60%	4.85	4.97	-2.42%

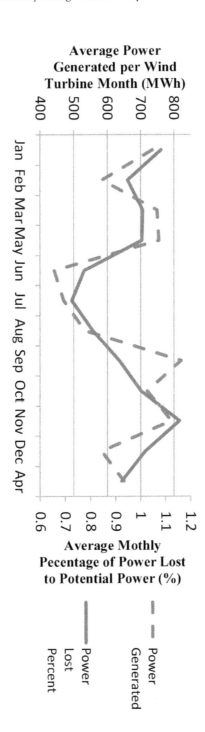

FIGURE 11: Salsburgh average monthly power generated and percentage of potential power lost due to failures.

The mean results of the MCMC analysis and equation 3 are shown in Table 4. The percentage difference between the MCMC model mean results and the static model is shown in the difference columns. The revenue of each potential site is estimated simply by using equation 3. As shown in Table 4 the revenue is over estimated compared to what is calculated by the MCMC model in all three sites. The largest difference is Salsburgh which is calculated by the model to have a reduced availability from the assumed value used in equation 3. The difference in revenue is also due to the wind turbine failing, according to the model, at times of the year when the wind is at its strongest.

The O&M cost estimated by the MCMC model is greater for each site than that calculated for the static model. The proportion of failures in the Salsburgh simulation which lead to downtimes of greater than 48 hours is 11.81 %, compared to 8.55 % of failures in the reliability data. This redistribution in severity of failures, along with a higher overall failure rate, accounts for the 57.84 % increase in projected O&M cost per wind turbine year. The static model does not account for this in its calculation.

The net income of each potential site is estimated using equation 5 and the operational output of the MCMC model. Comparing the results of the two methods, the difference between them shows that by neglecting the effect of wind speed dependent failure rates, the net income of the three sites are over estimated. Salsburgh is the site with the highest difference between the model and the non-model estimates, this is because of its greater average wind speed which increases the cost of O&M. However Prestwick—which has a relatively low wind speed—still shows a difference of 2.00 % in net income when comparing the methods. For a moderately sized wind farm of 50 wind turbines, this 2.00 % difference would equate to £4.5 million in lost revenue over twenty years.

There is a small difference in revenue shown in Table 4 for the three sites, calculated using both methods. This is due to the effects of the seasonal wind speeds on wind turbine reliability and productivity. Figure 11 shows the average power generated per month and the percentage of power lost on average each month because of component failures and resultant downtimes from the mean results of the MCMC simulation. As shown in Figure 11 the percentage of power lost to potential power increases in the winter months and decreases in the summer months. This seasonal trend cannot be modelled using equations 3- 5. The MCMC model takes account of this and any change in availability and the result is a difference of 0.72 %, 0.13 % and 0.12 % in revenue for Salsburgh, Prestwick and Leuchars respectively.

If a developer decided to build a wind farm at Salsburgh and chose take out a warranty with the original equipment manufacturer (OEM), they would benefit if the OEM did not account for the wind speed dependent failure rates. However this may mean that the OEM would recoup their costs at Salsburgh by increasing the price of the warranty on other sites. For a developer at a site with a less productive resource which causes fewer failures, such as Prestwick or Leuchars, this resultant higher O&M cost would make the investment less economical. However if the OEM factored the effects of individual site wind speed into their warranty, the costs to the operator would better reflect the reliability and the risk of their site.

Banks and other organisations that cover the operator's initial investment would also benefit from factoring in the impact of wind speed on reliability. For sites with high wind speeds, such as Salsburgh, there are significant differences between net incomes calculated when accounting for the impact of wind speed on reliability and when using static failure rates. According to Table 4, a wind farm sited at Salsburgh would generate 5.07 % less income than expected according to equation 5.

This is of course assuming a static O&M cost. In reality the cost of O&M would rise as the wind turbine and its components follow a bathtub curve and become less reliable and more expensive to maintain [22]. More work must be undertaken to understand how the wind speed effects wind turbines of different ages. As described in section 2, the wind turbines in this analysis are between 3 and 6 years old. The reliability data therefore shows a period of time during which the wind turbines would be expected to be very reliable and be operating at the bottom of the bath-tub curve [22].

3.4 CONCLUSION

This paper has demonstrated a method for assessing the reliability of a potential wind farm site by applying wind speed dependent failure rates calculated from a multi megawatt wind turbine reliability dataset and onsite met mast data. The advantages of using this method, as opposed to using static component failure rates, is that more informative O&M costs can be calculated and the effect of seasonal changes on wind turbine operation can be accounted for.

The site which will produce the most electricity is Salsburgh, which also has the strongest wind resource and the highest predicted failure rate. The model estimated an O&M cost which was 57.84 % higher than expected. The effect of more frequent and longer downtimes in the winter also reduced the estimated revenue by 0.72 %. The effect on the net income was a reduction of £460,000 the lifespan of the wind farm.

Future research will assess the impact of the wind speed on the reliability of wind turbines of different ages. Currently the results only describes the impact of the wind speed on one model of wind turbine—to make the model more useful, data from other wind turbine models will need to be used to calibrate the model. From this site specific model, more detailed assessments could be made including spares optimisation based on seasonal reliability. To properly assess the reliability and economic viability of a site, it is therefore important that the impact of the wind speed on the reliability of the components is assessed. The research in this paper will be of particular interest to operators and wind turbine manufacturers.

REFERENCES

1. P. J. Tavner, H. Long, and Y. Feng, "Early experiences with UK round 1 offshore wind farms," Proc. ICE - Energy, vol. 163, no. 4, pp. 167–181, Nov. 2010.
2. F. Besnard, K. Fischer, and L. Bertling, "Reliability-Centred Asset Maintenance—A step towards enhanced reliability, availability, and profitability of wind power plants," in EEE PES Conference on Innovative Smart Grid Technologies Europe, 2010, pp. 1–8.
3. L. Bertling, R. Eriksson, and R. N. Allan, "Relation between preventive maintenance and reliability," in IEEE Porto Power Tech Conference, 2001, vol 4.
4. G. J. W. Van Bussel and a R. Henderson, "State of the Art and Technology Trends for Offshore Wind Energy : Operation and Maintenance Issues," Offshore Conroe TX, pp. 10–12, 2001.
5. J. Phillips, P. Reynolds, L. Gosden, G. Hemmingsen, I. Mcdonald, G. Mackay, W. Hines, P. O. Repower, J. Brown, and J. Beresford, "A Guide to UK Offshore Wind Operations and Maintenance." Scottish enterpirse and the Crown Estate.

6. M. Wilkinson, G. L. G. Hassan, S. Vincent, S. Lane, B. Bs, T. Van Delft, and K. Harman, "The Effect of Environmental Parameters on Wind Turbine Reliability," in EWEA 2012 Copenhagen, Denmark, 2012.

7. B. Hahn, "Zeitlicher Zusammenhang von Schadenshäufigkeit und Windgeschwindigkeit," in FGW-Workshop Einflub der Witterung auf Windenergieanlagen, 1997.

8. P. J. Tavner, D. M. Greenwood, M. W. G. Whittle, R. Gindele, S. Faulstich, and B. Hahn, "Study of weather and location effects on wind turbine," Wind Energy, vol 16 no. 2, pp. 175–187, 2013.

9. G. Wilson and D. McMillan, "Modelling the Impact of the Environment on Offshore Wind Turbine Failure Rates," in EWEA Offshore, Frankfurt, Germany, 2013.

10. G. Wilson and D. McMillan, "Quantifying the Impact of Wind Speed on Wind Turbine Component Failure Rates," in EWEA 2014, Barcelona, Spain, 2014.

11. S. Faulstich, B. Hahn, P. Lyding, and P. Tavner, "Reliability of offshore turbines—identifying risks by onshore experience." EWEA Offshore, Stockholm, Sweden. 2009.

12. F. Castro Sayas and R. N. Allan, "Generation availability assessment of wind farms," IEE Proc. Gener. Transm. Distrib., vol. 143, no. 5, p. 507, 1996.

13. D. McMillan and G. W. Ault, "Quantification of Condition Monitoring Benefit for offshore wind turbines," Wind Eng., vol. 31, no. 4, 2007.

14. F. Besnard and L. Bertling, "An Approach for Condition-Based Maintenance Optimization Applied to Wind Turbine Blades," IEEE Trans. Sustain. Energy, vol. 1, no. 2, pp. 77–83, 2010.

15. R. Billington and R. N. Allan, Reliability Evaluation of engineering Systems: Concepts and Techniques, 1st ed. New York: Plenum Press, 1983.

16. D. McMillan and G. W. Ault, "Techno-economic comparison of operational aspects for direct drive and gearboxdriven wind turbines," IEEE Trans. Energy Convers., vol. 25, no. 1, pp. 191–198, 2010.

17. Met Office, "Met Office Integrated Data Archive System Land and Marine Surface Stations Data," 14-Oct-2013. [Online]. Available: http://badc.nerc.ac.uk/view/badc. nerc.ac.uk__ATOM__dataent_ukmo-midas.

18. T. Burton, D. Sharpe, N. Jenkins, and E. Bossanyi, Wind Energy Handbook, 1st ed. Chichester: John Wiley and Sons Inc, 2008, p. 19.

19. Mott MacDonald, "UK Electricity Generation Costs Update," 2010.

20. Ofgem, "Renewables Obligation: Guidance for Generators," London, 2013.

21. P. E. Morthorst and S. Awerbuch, "The Economics of Wind Energy," EWEA. 2009.

22. S. Faulstich, B. Hahn, and P. J. Tavner, "Wind turbine downtime and its importance for offshore deployment," Wind Energy, vol. 14, no. 3, pp. 327–337, 2011.

(WRF) model [10] so as to accurately simulate the wind conditions and to reproduce the stability effects on the wind profile over the MABL in the North Sea. The present work describes the verification of the WRF model, by comparing the model results with that of other modelling studies (e.g., [8, 11]) and with high quality observations recorded at the FINO1 offshore platform in the North Sea. For this verification process, several WRF modelling simulations have been performed during March 2005. In particular, different horizontal resolutions, PBL parameterizations, initial and boundary conditions, and nesting options were tested. The high probability (20%) of occurrence of stable to very-stable atmospheric stratification situations in the spring and early summer at the FINO1 platform and the impact of the high stability on the wind profile increases the need to focus on the stable atmospheric conditions and include them in our WRF model configuration and mesoscale methodology [3, 5]. A time period during which all the atmospheric stability conditions are observed are also investigated in this study.

A description of the verification process including information about the data sources and the statistics used for analysis is given in Section 2.1. Section 2.2 focuses on the WRF model setup and the process followed for testing its performance. Description of the synoptic conditions during the stable period in March 2005 and presentation of the WRF results for this case study with a number of key derived statistics are provided in Section 3. In the same section, presentation of the WRF results for the whole March 2005 with statistical metrics and error analysis is also given. Conclusions follow in Section 4.

4.2 MATERIALS AND METHODS

4.2.1 VERIFICATION PROCESS

Two main steps of the verification process are the definition of the data sources and the definition of the verification statistics to be produced by the analysis.

TABLE 1: Meteorological parameters with their associated heights and the sensor types including their accuracy [6] at the FINO1 offshore platform.

Variable	Heights (m) LAT	Sensor type (accuracy)
Wind speed (m/s) (cup anemometer)	34, 41.5, 51.5, 61.5, 71.5, 81.5, 91.5, 104.5	Vector A100LK-WR-PC3 (±0.01 m/s)
Wind direction (degree)	41.5, 51.5, 61.5, 71.5, 81.5, 91.5	Thies wind vane classic (±1°)
Air temperature (°C)	30, 41.5, 52, 72, 101	Pt-100 (±0.1 K at 0°C)
Relative humidity (%)	34.5, 90	Hydrometer, Thies (±3% RH)
Air pressure (hPa)	22.5	Barometer, Vaisala (±0.03 hPa)

4.2.1.1 OBSERVATIONAL DATA: FINO1 MEASUREMENTS

The FINO1 offshore platform in Southern North Sea is located 45 km north of the Borkum island (latitude: 54.0°N and longitude: 6.35°E) and performs multilevel measurements of wind speed, wind direction, air temperature, relative humidity, and air pressure since 2004. The height of the measurement mast is about 100 m above mean sea level (MSL). Three ultrasonic instruments (of 10 Hz temporal resolution) are located at 41.5 m, 61.5 m, and 81.5 m height on northwesterly oriented booms. In addition, eight cup anemometers with the lower resolution of 1 Hz are installed at different heights starting from 34 m upto about 100 m (every 10 m) on booms mounted in southeast direction of the meteorological mast (e.g., [8, 11]).

A list of parameters used for the analysis of the meteorological conditions and for the comparison with the WRF model results is shown in Table 1. In Table 1, we also provide the heights of the recorded parameters and the sensor types including their accuracy according to Cañadillas [6]. For the WRF model validation we used data from the cup anemometers in order to have the best spatial coverage to describe the vertical profiles of wind speed and wind direction. According to Deutsches Windenergie-Institut (DEWI, German Wind Energy Institute) scientists at the RAVE 2012 conference [12], the sonic anemometer data were sparse and were associated with errors especially during the period 2004–2007. In general, from 2004 to 2011 the availability of the cup anemometers at the FINO1 platform was about 98%, while the availability of the sonic anemometers was approximately 83% [12].

4.2.1.2 DEFINE VERIFICATION STATISTICS

The following statistical metrics have been used in this study to verify the performance of the WRF model when compared with the FINO1 observations. More details on the verification statistics could be found in [13].

Bias or mean error (ME) is defined as the mean of the differences between the WRF simulated meteorological parameters and the FINO1 observations. In particular, the mean error is calculated for each hour of data and the time average (over the period we have considered to study) bias is then provided for each measuring height of the FINO1 platform.

Mean absolute error (MAE) is defined as the quantity used to measure how close the observed values are to the modelled ones. The MAE is given by

$$MAE = \frac{1}{n}\sum_{i=1}^{n}|x_i - y_i|$$

(1)

where $|x_i - y_i|$ is the absolute error with x_i and y_i to represent the modelled and the observed values at the FINO1 met mast, respectively. n is the sample size.

Standard Deviation (STD) of the ME is defined as the dispersion of the biased values around the mean value. A low standard deviation indicates that the data points tend to be very close to the mean; high standard deviation indicates that the data points are spread out over a large range of values. The STD uses the following formula:

$$\sigma = \sqrt{\frac{\sum_{i=1}^{n}(x_i - \bar{x})^2}{(n-1)}}$$

(2)

where σ is the standard deviation and x is the sample mean.

Root Mean Square Error (RMSE) is a frequently used measure of the difference between values predicted by a model and the values actually observed. It measures the average magnitude of the error and it is defined as the measure of the combined systematic error (bias) and random error (standard deviation). Therefore, the RMSE will only be small when both the variance and the bias of an estimator are small. The RMSE uses the following formula:

$$RMSE = \sqrt{\sigma^2 + \bar{x}^2}$$

(3)

where x is the sample mean and σ is the standard deviation.

Pearson correlation coefficient (R) is defined as the measure of the linear dependence between the WRF results and the FINO1 data, giving a value between +1 and −1 inclusive. It thus indicates the strength and direction of a linear relationship between these two variables. A value of 1 implies that a linear equation describes the relationship between WRF and the observations perfectly, with all data points lying on a line for which the WRF values increase as the data values increase. The correlation is −1 in case of a decreasing linear relationship and the values in between indicates the degree of linear relationship between the WRF model and the observations. The formula for the Pearson product moment correlation coefficient is

$$R = \frac{\sum_{i=1}^{n}(x_i - \bar{x})(y_i - \bar{y})}{\sqrt{\sum_{i=1}^{n}(x_i - \bar{x})^2(y_i - \bar{y})^2}}$$

(4)

where the x and y are, respectively, the sample means of the WRF results and the measurements.

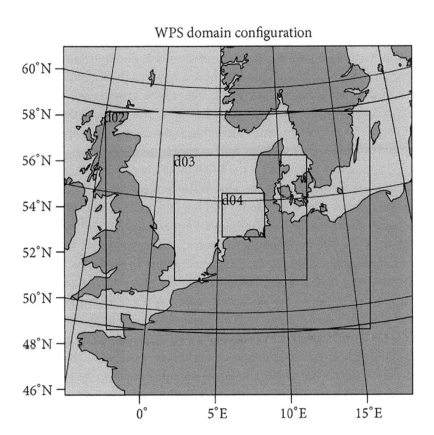

FIGURE 1: WRF modelling domain: 27 km (d01), 9 km (d02), 3 km (d03), and 1 km (d04) grid spacing.

4.2.2 MODEL AND SETUP

In this study, the numerical weather prediction (NWP) model of the National Centre for Atmospheric Research (NCAR): advanced WRF model, version 3.4 [10] was used. The model was run for this project on a Xeon X54 system at EDF R&D in France, with 96 CPUs. The WRF model is based on the fully compressible, nonhydrostatic Euler equations and for the purposes of this research the Lambert conformal projection was chosen. A third order Runge-Kutta (RK3) integration scheme and Arakawa C-grid staggering were used for temporal and spatial discretization, respectively. The modelling setup including the selected domains and the initial and boundary conditions as well as the physics schemes is described below. The strategy followed for the WRF modelling setup followed up to a certain point the strategy of previous WRF studies (e.g., [14, 15]).

Domain Setup. The WRF model was run in a series of two-way nested grids (centred in the FINO1 offshore platform at latitude: 54.0°N and longitude: 6.35°E). The horizontal grid spacing was refined by a factor of 3 through three nested domains until 1 km resolution. In particular, the WRF model was built over a parent domain (d01) with 67 × 65 horizontal grid points at 27 km, an intermediate nested domain (d02) of 9 km spatial resolution (151 × 121 grid points), and two innermost domain (d03) with 3 km spacing (217 × 205 grids) and 1 km spacing of 202 × 190 grids (see Figure 1). On the vertical coordinate, 88 vertical levels were used. The vertical resolution was 10 m upto 200 m height to accurately resolve the lower part of the MABL. Above 200 m, grid spacing is progressively stretched.

Initial and Boundary Conditions. The WRF model was used to refine the state of the atmosphere, especially the PBL, by downscaling both the global NCEP final analysis (FNL) data and the Era-Interim reanalysis data produced by the ECMWF. The NCEP FNL data have horizontal resolution of 1 × 1 degree (~100 km) and 52 model levels. The Era-Interim reanalysis project covers the period from 1979 to present and has a spectral T255 horizontal resolution (~79 km spacing on a N80 reduced Gaussian grid) and 60 vertical model levels [16]. The time step was set equal to 120 seconds to fulfil the Courant-Friedrichs-Lewy (CFL) condition for horizontal and vertical stability and the first 24 hours were discarded as spin-up time of the model.

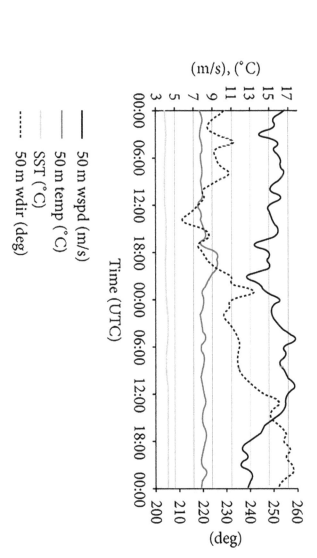

FIGURE 2: Temporal variation of the sea-surface temperature (SST), air temperature (°C), wind speed (m/s), and wind direction (degree) at 50 m height, as these were recorded at the FINO1 platform for the stable period 16–18 March 2005 at 00:00 UTC. Temperature and wind speed are represented in the left y-axis and the wind direction in the right x-axis of the plot.

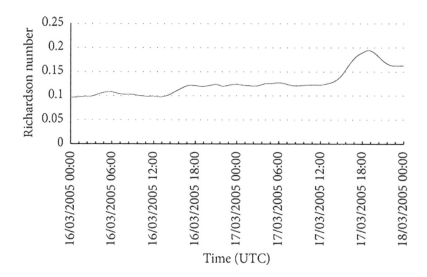

FIGURE 3: Temporal variation of the Richardson number, as this was calculated for the stable period 16–18 March 2005 at 00:00 UTC.

Topographic Inputs. For the WRF the topographic information was developed using the 20-category moderate resolution imaging spectroradiometer (MODIS) WRF terrain database. The 27 km domain was based on the 5 minutes (~9.25 km) global data, the 9 km domain was based on the 2 minutes (~3.70 km) data, and the 3 km and 1 km domains were based on the 30 seconds (~900 m) data.

Four-Dimensional Data Assimilation (FDDA). Nudging is one method of FDDA that is implemented in the WRF model. The WRF model supports three types of nudging: the three-dimensional upper-air and/or surface analysis, the observation, and the spectral nudging. Because the purpose of the stable case study was mainly to perform the sensitivity experiments rather than to keep the simulations close to the reanalysis and measurement data, the WRF model was run without analysis and observation nudging for this stable period. When the model was run for the whole March 2005, analysis nudging was configured to nudge temperature, water vapour, mixing ratio, and horizontal wind components on the outermost domain (d01) with time intervals of six hours. The strategy followed for analysis nudging was based on previous studies [17].

Physics Schemes. Microphysics was modelled using the new Thompson scheme with ice, snow, and graupel processes suitable for high-resolution simulations. The rapid radiative transfer model (RRTM) (an accurate and widely used scheme using look-up tables for efficiency) and the Dudhia (a simple downward integration allowing for efficient cloud and clear-sky absorption and scattering) schemes were used for the longwave and shortwave radiation options, respectively. The Noah land surface model (LSM) was chosen to simulate soil moisture and temperature and canopy moisture. Finally, the cumulus physics was modelled with the Grell 3D scheme, which is an improved version of the Grell-Devenyi (GD) ensemble scheme that can be used on high resolution (in addition to coarser resolutions). However, the cumulus parameterization was turned off in the fine-grid spacing of 3 km and 1 km selected for these simulations. Theoretically, it is only valid for parent grid sizes greater than 9 km [10]. Finally, four PBL schemes were selected for this study. These include two turbulent kinetic energy (TKE) closure schemes, the Mellor-Yamada-Janjic (MYJ) PBL [18] and the Mellor-Yamada Nakanishi and Niino Level 2.5 (MYNN) PBL [19] and two first-order closure schemes, the Yonsei University (YSU) PBL [20] and the asymmetric convective model version 2 (ACM2) [21]. More details of the model, of the physics schemes and references, could be found in [10].

4.3 RESULTS AND DISCUSSION

4.3.1 STABLE PERIOD (16–18 MARCH 2005)

In this section, WRF model results are presented and compared with other modelling studies and with high quality observations recorded at the FINO1 offshore platform in the North Sea. The model validation is for stable atmospheric conditions during March 2005. According to Krogsæter [3] and Saint-Drenan et al. [5], at the FINO1 offshore platform stable atmospheric conditions are observed 20% of the time during spring and early summer. It is well known that the mesoscale models handle both high atmospheric stability and high instability with difficulty due to their PBL schemes. The stable MABL is very complex and its structure is more complicated and variable than the structure of the unstable/neutral MABL [22]. Stable atmospheric conditions are mainly observed during the night and it is well known that the nocturnal boundary layer is driven by two distinct processes: low turbulence and radiative cooling which both are very difficult to describe and to model. Therefore, making precise offshore wind resource maps under stable conditions is a great challenge. For the representative case of the stable MABL, the period from 16 to 18 March 2005 at 00:00 UTC was selected for analysis. Similar studies have been performed in the past for this particular time period and we could therefore compare our methodologies and results with that of other researches (e.g., [8, 11]).

During the stable period, the synoptic situation over the North Sea is determined by a low-pressure system over the Atlantic and a high-pressure system over the southern Europe, which later in the examined period shifts to the north-west [11]. In Figure 2, we present the temporal variation of the wind and temperature conditions at the FINO1 offshore platform during the stable period. As can be seen in Figure 2, at the beginning of the period very strong southwesterly (~210–240°) winds dominate the area, which later become westerlies (~270°) and slow down. It can be assumed that the MABL structure over the North Sea is influenced by large scale circulations, since no diurnal variation is observed in the air temperature and wind speed. According to Sušelj and Sood [11], the MABL structure over the North Sea is primarily established by the properties of advected air masses and has small or almost no diurnal cycle. For example, at the FINO1 platform mainly during the winter, the south-westerlies to westerlies advect warm air over the cold sea, resulting in a stable MABL, while north-westerlies to north-easterlies advect cold air, resulting in an unstable MABL. In general, we could characterise the period 16–18 March 2005 as a two-day stable period with the air temperature higher than the water temperature and with high wind speeds (see Figure 2). Actually, the FINO1 recordings gave even a 5°C difference between the air and the water temperature at the night of 16 March

2005, corresponding closely to stable conditions. Considering the air-water temperature difference is a very simple and pragmatic method of describing the stability conditions. In the current study, this method was shown to be sufficiently precise. In addition, the Richardson number has been also calculated to confirm the stability conditions.

To determine the atmospheric stability, the Richardson number has been calculated using the WRF model outputs. The Richardson number quantifies the respective contribution of the wind shear and the buoyancy to the production of turbulent kinetic energy [5] and the equation used in the computation of the Richardson number was

$$Ri = \frac{g}{\theta_1} \frac{(\theta_1 - \theta_2)(z_1 - z_2)}{(u_1 - u_2)^2} \tag{5}$$

where g is the gravitational acceleration, u_1, u_2, θ_1, and θ_2 are the wind speed and the virtual potential temperature at the height z_1 and z_2, respectively [23].

It is worth noting that negative Richardson numbers indicate unstable conditions; positive values indicate statically stable flows and values close to zero or zero are indicative of neutral conditions. As an example, the Richardson number is plotted in Figure 3 for the period 16–18 March 2005. The positive values of the Richardson number (varying from 0.1 to 0.2) confirm the statically stable flows that dominated during this particular time period.

4.3.1.1 SENSITIVITY TO HORIZONTAL RESOLUTION

Increasing the horizontal resolution in the mesoscale models increases the ability of the models to resolve major features of the topography and surface characteristics, such as the coastal boundaries and therefore produces more accurate wind climatologies [4]. Generally speaking, the higher the resolution of the simulation, the better the representation of the atmospheric processes we obtain [24], especially if the terrain is complex [4].

However, it is difficult to define a priori the grid spacing needed to achieve a desired level of accuracy. The sensitivity to horizontal resolutions is thus tested in this section to define optimum grid spacing. For the runs, the ERA-Interim reanalysis was selected to initialise the model and the YSU scheme to model the MABL.

To evaluate the performance of the different horizontal resolutions, four statistical parameters were used: mean absolute error (MAE), mean error (ME), standard deviation (STD) of the ME, and correlation coefficient (R). As an example, statistical metrics for the 100 m wind speed are shown in Table 2 for the 3 km and 1 km spacing. It seems that the 3 km horizontal resolution results in a slight reduction of MAE, bias, and STD and improves correlations with the observations compared to what has been simulated at the 1 km spacing. We concluded that the 3 km is the optimal resolution and that the finest 1 km spacing does not bring improvement in the WRF results worth considering. It seems that the parameterizations used in the WRF model impose a limit to the downscaling, beyond which there is a minor or any improvement of the model performance. According to Talbot et al. [25], it is beneficial to use very fine horizontal resolution (≤1 km) during stable atmospheric conditions and over heterogeneous surfaces when local properties are required or when resolving small-scale surface features is desirable. In this modelling study, the North Sea can be characterised as a homogeneous surface and therefore there is no advantage of using the computationally expensive 1 km spacing (as the model resolution increases more computer processing required). It is worth noting that similar results to the one presented in this section were obtained for all the vertical levels below 100 m.

4.3.1.2 SENSITIVITY TO INPUT DATA

The impact of using different reanalysis datasets on the initialisation of the WRF model is investigated in this section. Comparing the model results when the ERA-Interim and the NCEP data were used at the 3 km horizontal resolution and with the YSU PBL scheme, it is found that the simulated winds are better correlated with the FINO1 observations and have lower STD and MAE when the ERA-Interim dataset is used (see Table 3). On average over all the vertical levels the ERA-Interim dataset yields 0.91 m/s lower wind speeds than recorded, 1.17 m/s STD, and correlation between the model and the measurements equal 0.77.

TABLE 2: Statistical metrics (MAE in m/s, ME in m/s, STD of the ME in m/s, and R) for the 100 m wind speed, so as to evaluate the performance of different horizontal resolutions (3 km versus 1 km).

At 100 m	MAE (m/s)	ME (m/s)	STD (m/s)	R
1 km—ERA-I	2.45	−2.31	1.58	0.62
3 km—ERA-I	2.38	−2.27	1.47	0.66

TABLE 3: Statistical metrics (MAE in m/s, ME in m/s, STD of the ME in m/s, and R) for wind speed at 8 vertical levels (30–100 m), so as to evaluate the performance of different input datasets (NCEP versus ERA-Interim) at the 3 km WRF modelling domain.

	30 m	40 m	50 m	60 m	70 m	80 m	90 m	100 m	Average
NCEP FNL and YSU PBL									
MAE (m/s)	1.09	1.09	1.06	1.23	1.44	1.94	2.20	2.40	1.56
ME (m/s)	0.51	0.44	−0.25	−0.75	−1.1	−1.8	−2.07	−2.33	−0.92
STD (m/s)	1.35	1.37	1.39	1.41	1.42	1.42	1.42	1.57	1.42
R	0.63	0.62	0.62	0.62	0.62	0.62	0.63	0.59	0.62
ERA-I and YSU PBL									
MAE (m/s)	0.89	0.90	0.88	1.07	1.29	1.84	2.10	2.38	**1.42**
ME (m/s)	0.46	0.39	−0.23	−0.74	−1.09	−1.75	−2.05	−2.27	**−0.91**
STD (m/s)	1.01	1.05	1.10	1.14	1.18	1.21	1.23	1.47	**1.17**
R	0.81	0.80	0.79	0.78	0.77	0.76	0.75	0.66	**0.77**

Note that in both experiments with different input data an increased MAE with the elevation is observed and the WRF model is found to overestimate the winds below 40 m and underestimate them above that height. It is also observed that the correlations slightly decrease and the bias continues to increase as the altitude increases; this trend in the mean error and correlation coefficient with the increasing height is an indication that the current model configuration needs further improvement. It seems that this decreasing trend is more pronounced when the ERA-Interim data were used, but the average results (considering the average statistical metrics: correlation, bias, STD, and MAE) were closer to the observations with this dataset. As will be shown later, these reduced correlations with increasing height are a result of several factors including the input data and mainly the PBL scheme.

TABLE 4: A synopsis of the WRF model configuration used by EDF R&D for the stable period: 16–18 March 2005.

Simulation period	15–20/03/2005, 00:00 UTC
Model version	V3.4
Domains	4
Horizontal resolution	27, 9, 3 and 1 km
Grid sizes	67×65, 151×121, 217×205
and 202×190	
Vertical resolution	88 (eta levels)
Input data	ERA-Interim
Time step	120 s
Outputs frequency	180, 180, 60 and 60 minutes
Gridded analysis nudging	NO
Nesting	2-way nesting
Physics schemes	
Microphysics	New Thompson
Cumulus	Grell 3D
Shortwave	Dudhia
Longwave	RRTM
LSM	Noah
PBL	YSU, MYNN, MYJ, ACM2

4.3.1.3 SENSITIVITY TO PBL SCHEMES

It is well accepted that the accuracy of the simulated, by the mesoscale models, offshore winds is strongly affected by the PBL schemes [26]. Therefore, a way to improve the WRF performance at the lower levels of the atmosphere is to try different PBL parameterisations. The behaviour of four different PBL schemes (YSU, MYJ, MYNN, and ACM2) on the stable MABL is examined and the importance of defining the appropriate model setup for studying pure offshore wind conditions is noticed in this section. The PBL experiment was run four times; one run for each PBL scheme. A synopsis of the model configuration is given in Table 4.

Statistical metrics (MAE, ME, STD, and R) for the wind speed at heights from 30 m upto 100 m are presented in Table 5. Note that the MYNN PBL scheme predicts lower values of MAE and ME in all vertical levels than the other PBL schemes. It seems that the MYNN scheme is the one that yields to the highest degree of correctness mainly above 50 m and thus showing in average the best agreement with the FINO1 observations. It is also noticed that the bias between the WRF model and the measurements is lower and close to the surface for the YSU, MYJ, and ACM2 PBL runs, whereas the bias increases with altitude. The YSU run seems to correlate better with the FINO1 measurements than the other PBL schemes below 70 m, but above that height the ME significantly increases and lower values of correlation were found. On the other hand, the MYNN scheme performs almost with the same accuracy at all heights with low values of ME and STD. In particular, the MYNN scheme resulted (in average over all the heights) in 0.01 m/s higher wind speeds than measured, STD of 1.21 m/s, and correlation coefficient equal to 0.66 (see Table 5).

It is well known that the atmospheric stability has a significant effect on the vertical wind profile and on estimates of the wind resource at a given height [4]. A detailed analysis of the offshore vertical wind profile upto hub heights and above is important for a correct estimation of the MABL wind and stability conditions, wind resource, and power forecasting.

The time-averaged vertical profiles of wind speed, temperature, and wind direction for all the PBL runs are provided in Figure 4 for comparison with the FINO1 measurements. This kind of representation allows us to have some understanding about the behaviour of different PBL schemes in the lower levels of the MABL. The highly stable conditions observed during the examined period pose a particular challenge. As can be seen in Figure 4, the YSU scheme was in good agreement with the observed data at the 30 m height, but higher up underestimated the wind resource. This result highlights the importance of studying the whole MABL for offshore wind energy applications. Also, an increase in wind speed with increasing altitude is noticed at all PBL runs, but only the MYNN scheme produces wind speeds which are in exceptionally good agreement with the observations. It seems that the MYNN PBL scheme manages to model the correct amount of mixing in the boundary layer. The MYNN scheme is also shown to be the most consistent with the temperature measurements (approximately 1°C difference at 100 m height) and gives less than 10° difference in the wind direction at all vertical levels (see Figure 4).

TABLE 5: Statistical metrics (MAE in m/s, ME in m/s, STD of the ME in m/s, and R) for the wind speed at 8 vertical levels (30–100 m), so as to evaluate the performance of four PBL schemes (YSU, MYJ, ACM2, and MYNN) at the 3 km WRF modelling domain.

	30 m	40 m	50 m	60 m	70 m	80 m	90 m	100 m	Average
YSU PBL									
MAE (m/s)	0.89	0.90	0.88	1.07	1.29	1.84	2.10	2.38	1.42
ME (m/s)	0.46	0.39	-0.23	-0.74	-1.09	-1.75	-2.05	-2.27	-0.91
STD (m/s)	1.01	1.05	1.10	1.14	1.18	1.21	1.23	1.47	1.17
R	0.81	0.80	0.79	0.78	0.77	0.76	0.75	0.66	0.77
MYJ PBL									
MAE (m/s)	1.38	1.25	1.43	1.55	1.56	1.79	1.76	1.64	1.55
ME (m/s)	-1.03	-0.84	-1.12	-1.25	-1.23	-1.52	-1.46	-1.33	-1.22
STD (m/s)	1.24	1.23	1.24	1.27	1.32	1.36	1.39	1.43	1.31
R	0.68	0.67	0.67	0.66	0.64	0.63	0.63	0.65	0.65
ACM2 PBL									
MAE (m/s)	1.08	0.97	1.20	1.38	1.46	1.77	1.79	1.86	1.44
ME (m/s)	-0.74	-0.56	-0.91	-1.15	-1.23	-1.61	-1.64	-1.59	-1.18
STD (m/s)	1.13	1.13	1.15	1.17	1.18	1.20	1.21	1.41	1.20
R	0.67	0.66	0.65	0.63	0.61	0.61	0.60	0.56	0.62
MYNN PBL									
MAE (m/s)	0.91	0.92	0.88	0.91	0.93	1.00	0.99	1.03	0.95
ME (m/s)	0.12	0.38	0.16	0.01	0.03	-0.27	-0.21	-0.12	0.01
STD (m/s)	1.22	1.21	1.20	1.21	1.23	1.23	1.26	1.28	1.23
R	0.67	0.67	0.66	0.66	0.64	0.64	0.63	0.67	0.66

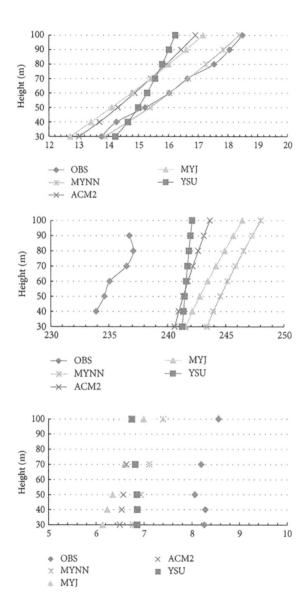

FIGURE 4: Time-averaged (16–18 March 2005) vertical profiles upto 100 m ASL for (a) wind speed (m/s), (b) wind direction (degree), and (c) temperature (°C). The model results of different PBL schemes are compared with the FINO1 measurements.

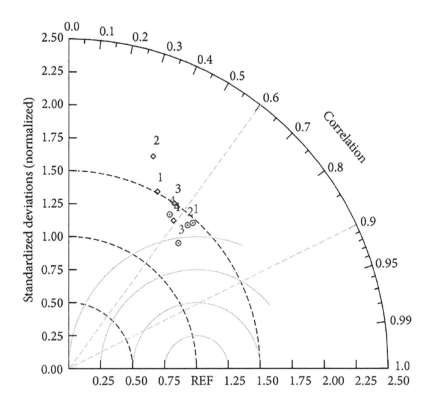

FIGURE 5: Taylor diagram displaying a statistical comparison with four WRF model estimates (YSU: 1, MYJ: 2, MYNN: 3, and ACM2: 4) of the wind speed during the stable period 16–18 March 2005. Red is the EDF statistics and blue is the Muñoz-Esparza et al. [8] statistics.

Comparison with Other Modelling Studies. Muñoz-Esparza et al. [8] in their WRF modelling study for the stable period 16–18 March 2005 compared six different PBL schemes. Their model configuration is presented in Table 6.

TABLE 6: A synopsis of the WRF model configuration used by Muñoz-Esparza et al. [8] for the stable period: 16–18 March 2005.

Simulation period	14–18/03/2005, 00:00 UTC
Model version	V3.2
Domains	4
Horizontal resolution	27, 9, 3 and 1 km
Vertical resolution	46
Input data	NCEP
Gridded analysis nudging	NO
Nesting	2-way nesting
Physics schemes	
Microphysics	WSM3
Cumulus	Kain-Fritsch
Shortwave	Dudhia
Longwave	RRTM
LSM	Noah
PBL	YSU, MYJ, MYNN, ACM2,
QNSE, BouLac	

Statistical metrics (ME, RMSE, and R) of the 100 m wind speed for the common PBL runs, as these were calculated by Muñoz-Esparza et al. [8] and by EDF, are shown in Table 7. Note that all PBL runs of this study result in a decreased bias and RMSE as well as in an increased correlation coefficient. The differences between EDF's and Muñoz-Esparza et al.'s [8] model setup, such as the input data, the number of the vertical levels, and in general the combination of physics schemes (e.g., microphysics and cumulus), may be a reason for the better accuracy in this study. For example, EDF initialised the model with the ERA-Interim data instead of the NCEP reanalysis and used 88 vertical levels instead of the 46 used by and

Muñoz-Esparza et al. [8]. It is well expected that part of the problem with the simulation of stable conditions in the MABL is having enough vertical levels. A reason is the fact that a stable layer acts as an effective barrier to mixing and if the model layers at the lower levels of the MABL are widely spaced, the simulated barrier may be too weak [4].

TABLE 7: Muñoz-Esparza et al. [8] versus EDF statistical metrics (ME in m/s, RMSE in m/s, and R) for the 100 m wind speed, so as to evaluate the WRF performance with different PBL schemes.

PBL schemes	[8]			EDF		
	ME (m/s)	RMSE (m/s)	R (—)	ME (m/s)	RMSE (m/s)	R (—)
YSU	−2.83	3.21	0.46	−2.27	2.70	0.66
MYJ	−1.93	2.60	0.38	−1.33	1.94	0.65
QNSE	−1.23	1.90	0.53	—	—	—
MYNN	−0.44	1.56	0.55	−0.12	1.28	0.67
ACM2	−1.61	2.13	0.59	−1.59	2.11	0.56
BouLac	−3.08	3.42	0.46	—	—	—

Another important point seen in Table 7 is that both Muñoz-Esparza et al. [8] and EDF concluded that the best model performance was observed when the MYNN PBL scheme was selected and that the WRF model tends to underestimate the wind speed at the hub height. The similarities between the PBL experiments of Muñoz-Esparza et al. [8] and EDF are quantified in terms of their correlation, their RMSE, and their STD by using the Taylor diagram [27]. In particular, Figure 5 is a Taylor diagram which shows the relative skill with which the PBL runs of Muñoz-Esparza et al. [8] and of EDF simulate the wind speeds during the stable period 16–18 March 2005. Statistics for four PBL runs (YSU: 1, MYJ: 2, MYNN: 3, and ACM2: 4) are used in the plot and the position of each number appearing on the plot quantifies how closely that run matches observations. If for example, run number 3 (MYNN PBL run) is considered, its pattern correlation with observations is about 0.67 for the EDF MYNN PBL experiment (red colour) and 0.55 for the Muñoz-Esparza et al. [8] MYNN PBL run (blue colour). It seems that the simulated patterns between EDF and [8] that agree well with each other are for the ACM2 PBL run (number 4) and the pattern that has relatively high correlation and low RMSE are for the EDF MYNN PBL run (red-number 3).

4.3.1.4 SENSITIVITY TO NESTING OPTIONS

In the WRF model, the horizontal nesting allows resolution to be focused over a particular region by introducing a finer grid (or grids) into the simulation [10]. Therefore, a nested run can be defined as a finer grid resolution model run in which multiple domains (of different horizontal resolutions) can be run either independently as separate model simulations or simultaneously.

The WRF model supports one-way and two-way grid nesting techniques, where one-way and two-way refer to how a coarse and a fine domain interact. In both the one-way and two-way nesting options, the initial and lateral boundary conditions for the nest domain are provided by the parent domain, together with input from higher resolution terrestrial fields and masked surface fields [10]. In the one-way nesting option, information exchange between the coarse domain and the nest is strictly downscale, which means that the nest does not impact the parent domain's solution. In the two-way nest integration, the exchange of information between the coarse domain and the nest goes both ways. The nest's solution also impacts the parent's solution.

To this point, the best model performance was achieved with the following combination: Era-Interim as input and boundary data, 3 km as the optimal horizontal resolution, and the MYNN PBL scheme to simulate the MABL. In this section, both the one-way and two-way nesting options with the aforementioned model combination were tested so as to investigate how the different nesting options can affect the WRF performance.

It seems that the differences between the two-way and one-way nesting options are small. However, it can be concluded that the two-way nest integration produces on average wind speeds closer to the FINO1 measurements (see Table 8). When the two-nesting option was selected, the average over all heights ME was 0.01 m/s and the STD of the ME was 1.23 m/s (instead of 0.06 m/s ME and 1.32 m/s STD for the one-way nesting option). As can be also seen in Table 8, the WRF model resulted in smaller MAE (averaged over all the heights) and correlated better with the observations when the two-way nesting option was selected. It is thus concluded that when the exchange of information between the coarse domain and the nest goes both ways, it results in better model performance.

TABLE 8: Statistical metrics (MAE in m/s, ME in m/s, STD of the ME in m/s, and) for the wind speed at 8 vertical levels (30–100 m), so as to evaluate the performance of one- and two-way nesting options at the 3 km WRF modelling domain.

	30 m	40 m	50 m	60 m	70 m	80 m	90 m	100 m	Average
MYNN and 2-way									
MAE (m/s)	**0.91**	**0.92**	**0.88**	**0.91**	**0.93**	**1.00**	**0.99**	**1.03**	**0.95**
ME (m/s)	0.12	0.38	0.16	0.01	0.03	−0.27	−0.21	−0.12	**0.01**
STD (m/s)	1.22	1.21	1.20	1.21	1.23	1.23	1.26	1.28	**1.23**
R	0.67	0.67	0.66	0.66	0.64	0.64	0.63	0.67	**0.66**
MYNN and 1-way									
MAE (m/s)	0.97	1.00	0.98	1.00	1.02	1.07	1.08	1.11	1.03
ME (m/s)	0.17	0.43	0.21	0.05	0.08	−0.22	−0.17	−0.07	0.06
STD (m/s)	1.28	1.28	1.28	1.30	1.33	1.34	1.36	1.40	1.32
R	0.64	0.63	0.62	0.61	0.59	0.59	0.58	0.62	0.61

4.3.1.5 SUMMARY

The validation of the WRF model at the FINO1 met mast during the stable case study (16–18 March 2005) resulted in the following conclusions in terms of model configuration.

1. Selecting the finest horizontal resolution of 1 km instead of the 3 km does not bring improvement in the WRF results worth considering; it was very time consuming in terms of computational time.
2. Using the ERA-Interim reanalysis data instead of the NCEP FNL data allows reducing the bias between simulated and measured winds.

3. Changing PBL schemes has a strong impact on the model results. The MYNN PBL scheme was in the best agreement with the observations (confirmed by other modelling studies as well).
4. A small improvement in the model performance was observed when the two-way nesting option was used.

4.3.2 MARCH 2005

Accuracy verification at the FINO1 platform indicated that the WRF methodology developed for the stable period (16–18 March 2005) results in more accurate wind simulation than any other simulation in the study of Muñoz-Esparza et al. [8], though there are similarities between the studies in terms of findings. But, is the concluded WRF model configuration appropriate for other stability conditions (e.g., neutral and unstable) and for longer time periods during which all atmospheric stability conditions are observed?

To answer this question, the WRF model is validated in this section for the whole of March 2005 in which neutral, unstable, and stable atmospheric conditions are observed. To determine the atmospheric stability at the FINO1 met mast for March 2005, the Richardson number was calculated. It was found out that 35% of the time during March 2005 stable atmospheric conditions were dominant at the FINO1 offshore platform, while 65% of the time unstable and neutral conditions were observed.

4.3.2.1 SENSITIVITY TO PBL SCHEMES

The WRF model is tested with the same four PBL schemes (YSU, MYJ, ACM2, and MYNN) used for the stable scenario. The monthly bias and standard deviation for each height of the FINO1 offshore platform is given in Figure 6 for each PBL run. In Figure 6, we can clearly observe that the MYNN PBL run gives the lowest bias at all vertical heights and the second lowest standard deviation. The biases of the YSU and ACM2 PBL schemes tend to increase as the height increases. As before for the stable case study, we concluded that the MYNN scheme is the one that yields to the highest degree of correctness with the FINO1 observational data. In particular, the MYNN scheme resulted in average over all the heights in 0.01 m/s lower wind speeds, STD of 1.71 m/s, and very high correlation coefficient equal to 0.93 (see Table 9).

TABLE 9: For March 2005, statistical metrics (ME in m/s, STD of the ME in m/s, and R) for wind speed [averaged over all the heights: 30 to 100 m)], so as to evaluate the performance of four PBL schemes (YSU, MYJ, ACM2, and MYNN) in the WRF model.

PBL	YSU	MYJ	ACM2	MYNN
ME (m/s)	−0.43	−0.26	−0.48	−0.01
STD (m/s)	1.75	1.75	1.69	1.71
R	0.92	0.92	0.92	0.93

In Figure 7, a map plot of the average 100 m wind speed and wind direction, as these were simulated during March 2005 from the MYNN PBL run over the 3 km WRF modelling domain, is provided. At the FINO1 mast (latitude: 54.0°N and longitude: 6.35°E) and the areas nearby the platform over the North Sea, the WRF model simulated hub height-wind speeds of the order of 11.2 m/s and westerly to southwesterly direction. In Figure 8, the wind roses are shown for March 2005 from both the FINO1 observations and all the WRF PBL simulations. It seems that the model in every PBL simulation overestimates the occurrence of wind speed in the range 12–16 m/s and results in a small veering of the winds in the clockwise direction between 0° and 180°.

4.3.2.2 ERROR ANALYSIS AND Q-Q PLOTS

In Figure 9, the Q-Q plots of the 100m wind speed for March 2005 display a quantile-quantile plot of two samples: modelled vs. observed. Each blue point in the Q-Q plots corresponds to one of the quantiles of the distribution the modelled wind follows and is plotted against the same quantile of the FINO1 observations' distribution. If the samples come from the same distribution, the plot will be linear and the blue points will lie on the 45° line y=x.

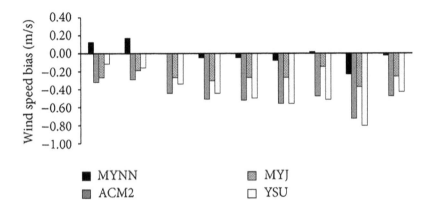

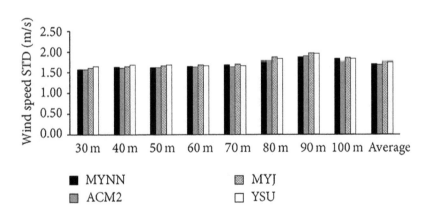

FIGURE 6: Monthly (March 2005) bias and standard deviation for wind speed at 8 vertical levels (30–100 m) and their average for the four PBL runs (YSU, MYJ, ACM2, and MYNN).

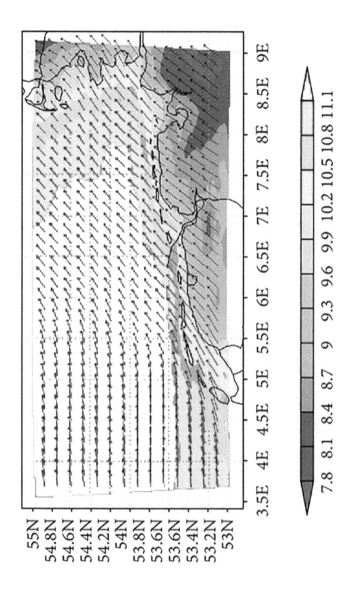

FIGURE 7: Average wind speed (m/s) and wind direction at 100 m height over the 3 km WRF modelling domain during March 2005 when the MYNN PBL scheme was selected.

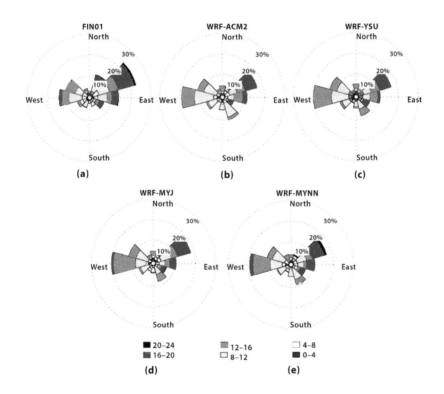

FIGURE 8: Wind roses based on March 2005 data from the FINO1 measurements and the WRF PBL runs (ACM2, YSU, MYJ, and MYNN) at 100 m height.

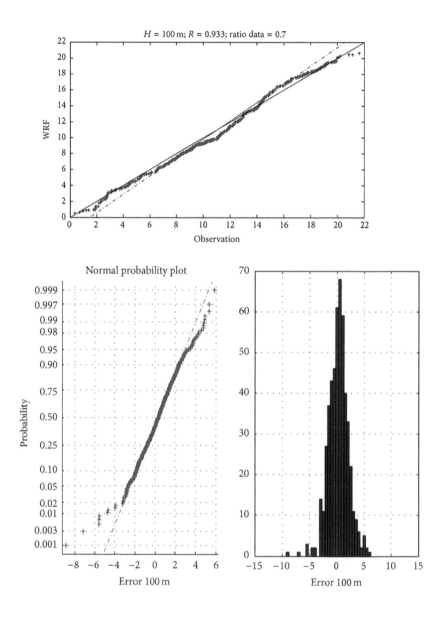

FIGURE 9: (a) Q-Q plot: the WRF MYNN PBL run (y-axis) against the FINO1 100 m wind speed distribution (x-axis). (b) Normal distribution against the 100 m wind speed ME and (c) histogram of the wind speed ME.

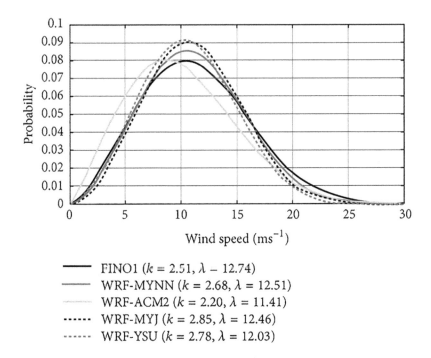

FIGURE 10: Weibull distribution based on March 2005 data from the FINO1 measurements and the WRF PBL runs (ACM2, YSU, MYJ, and MYNN) at 100 m height.

As can be seen in Figure 9, the two distributions (the one modelled with the MYNN PBL scheme versus the observed one at FINO1) being compared are almost identical. The blue points in the Q-Q plot are closely following the 45° line y=x with high correlation coefficient equal to 0.93 (see Table 9). The hourly observed and modelled wind speeds for March 2005 follow the Weibull distribution and upon closer inspection of Figure 10 the Weibull distribution resulted from the MYNN PBL run is the one that fits very well with the FINO1 observed distribution at the 100 m height. In particular, the shape and scale parameters of the Weibull distribution, as these were calculated with the FINO1 measurements, were 2.51 and 12.74, respectively. The shape parameter modelled by the WRF model varied from 2.20 to 2.85 and the scale parameter ranged from 11.41 to 12.51 for the different PBL schemes used. However, only the MYNN PBL scheme resulted in shape and scale parameters closer to the one observed at the offshore platform (see Figure 10).

In order to further understand the performance of the WRF model and show the variation of the model results, in this section an error analysis is also performed. In general, the errors are assumed to follow normal distributions about their mean value and so the standard deviation is the measurement of the uncertainty. If it turns out that the random errors in the process are not normally distributed, then any inferences made about the process may be incorrect. In Figure 9, we provide a normal probability plot of the ME of the 100 m wind speed. The plot includes a reference line indicating that the ME of the wind speed (blue points) closely follows the normal distribution. Figure 9 also shows a histogram of the wind speed mean error. In the y-axis of this histogram we have the number of records with a particular velocity error. The blue bars of the histogram show the relative frequency with which each wind speed error (which is shown along the x-axis) occurs. Clearly the ME follows the normal distribution and the mean is very close to zero, which is an indication that the mean error is unbiased.

4.4 CONCLUSIONS

The main goals of the present study were to achieve a better understanding of the offshore wind conditions, to accurately simulate the wind flow and atmospheric stability occurring offshore over the North Sea, and to develop a mesoscale modelling approach that could be used for more accurate WRA. In order to achieve all these goals, this paper examined the sensitivity of the performance of the WRF model to the use of different initial and boundary conditions, horizontal resolutions, and PBL schemes at the FINO1 offshore platform.

The WRF model methodology developed in this study appeared to be a very valuable tool for the determination of the offshore wind and stability conditions in the North Sea. It resulted in more accurate wind speed fields than the one simulated in previous studies, though there were similar findings. It was concluded that the mesoscale models must be adaptive in order to increase their accuracy. For example, the atmospheric stability on the MABL (especially the stable atmospheric conditions), the topographic features in order to adjust the horizontal resolution accordingly, and the input datasets are needed to be taken into account.

It was concluded that the 3 km spacing is an optimal horizontal resolution for making precise offshore wind resource maps. Also, using the ERA-Interim reanalysis data instead of the NCEP FNL data allows us to reduce the bias between simulated and measured winds. Finally, it was found that changing PBL schemes has a strong impact on the model results. In particular, for March 2005 the MYNN PBL run yields in the best agreement with the observations with 0.01 m/s bias, standard deviation of 1.71 m/s, and correlation of 0.93 (averaged over all the vertical levels).

In future studies, in order to increase the reliability of the wind simulations and forecasting, it is necessary to expand the simulation period as long as possible.

REFERENCES

1. O. Krogsæter, J. Reuder, and G. Hauge, "WRF and the marine planetary boundary layer," in EWEA, pp. 1–6, Brussels, Belgium, 2011.
2. A. C. Fitch, J. B. Olson, J. K. Lundquist et al., "Local and Mesoscale Impacts of Wind Farms as Parameterized in a Mesoscale NWP Model," Monthly Weather Review, vol. 140, no. 9, pp. 3017–3038, 2012.
3. O. Krogsæter, "One year modelling and observations of the Marine Atmospheric Boundary Layer (MABL)," in NORCOWE, pp. 10–13, 2011.

4.   M. Brower, J. W. Zack, B. Bailey, M. N. Schwartz, and D. L. Elliott, "Mesoscale modelling as a tool for wind resource assessment and mapping," in Proceedings of the 14th Conference on Applied Climatology, pp. 1–7, American Meteorological Society, 2004.

5.   Y.-M. Saint-Drenan, S. Hagemann, L. Bernhard, and J. Tambke, "Uncertainty in the vertical extrapolation of the wind speed due to the measurement accuracy of the temperature difference," in European Wind Energy Conference & Exhibition (EWEC '09), pp. 1–5, Marseille, France, March 2009.

6.   B. Cañadillas, D. Muñoz-esparza, and T. Neumann, "Fluxes estimation and the derivation of the atmospheric stability at the offshore mast FINO1," in EWEA Offshore, pp. 1–10, Amsterdam, The Netherlands, 2011.

7.   D. Muñoz-Esparza and B. Cañadillas, "Forecasting the Diabatic Offshore Wind Profile at FINO1 with the WRF Mesoscale Model," DEWI Magazine, vol. 40, pp. 73–79, 2012.

8.   D. Muñoz-Esparza, J. Beeck Van, and B. Cañadillas, "Impact of turbulence modeling on the performance of WRF model for offshore short-term wind energy applications," in Proceedings of the 13th International Conference on Wind Engineering, pp. 1–8, 2011.

9.   M. Garcia-Diez, J. Fernandez, L. Fita, and C. Yague, "Seasonal dependence of WRF model biases and sensitivity to PBL schemes over Europe," Quartely Journal of Royal Meteorological Society, vol. 139, pp. 501–514, 2013.

10. W. Skamarock, J. Klemp, J. Dudhia et al., "A description of the advanced research WRF version 3," NCAR Technical Note 475+STR, 2008.

11. K. Sušelj and A. Sood, "Improving the Mellor-Yamada-Janjić Parameterization for wind conditions in the marine planetary boundary layer," Boundary-Layer Meteorology, vol. 136, no. 2, pp. 301–324, 2010.

12. F. Kinder, "Meteorological conditions at FINO1 in the vicinity of Alpha Ventus," in RAVE Conference, Bremerhaven, Germany, 2012.

13. J. M. Potts, "Basic concepts," in Forecast Verification: A Practitioner's Guide in Atmospheric Science, I. T. Jolliffe and D. B. Stephenson, Eds., pp. 11–29, John Wiley & Sons, New York, NY, USA, 2nd edition, 2011.

14. E. M. Giannakopoulou and R. Toumi, "The Persian Gulf summertime low-level jet over sloping terrain," Quarterly Journal of the Royal Meteorological Society, vol. 138, no. 662, pp. 145–157, 2012.

15. E. M. Giannakopoulou and R. Toumi, "Impacts of the Nile Delta land-use on the local climate," Atmospheric Science Letters, vol. 13, no. 3, pp. 208–215, 2012.

16. D. P. Dee, S. M. Uppala, A. J. Simmons et al., "The ERA-Interim reanalysis: configuration and performance of the data assimilation system," Quarterly Journal of the Royal Meteorological Society, vol. 137, no. 656, pp. 553–597, 2011.

17. P. Liu, A. P. Tsimpidi, Y. Hu, B. Stone, A. G. Russell, and A. Nenes, "Differences between downscaling with spectral and grid nudging using WRF," Atmospheric Chemistry and Physics, vol. 12, no. 8, pp. 3601–3610, 2012.

18. G. L. Mellor and T. Yamada, "Development of a turbulence closure model for geophysical fluid problems," Reviews of Geophysics & Space Physics, vol. 20, no. 4, pp. 851–875, 1982.

19. M. Nakanishi and H. Niino, "An improved Mellor-Yamada Level-3 model with condensation physics: its design and verification," Boundary-Layer Meteorology, vol. 112, no. 1, pp. 1–31, 2004.

20. S.-Y. Hong, Y. Noh, and J. Dudhia, "A new vertical diffusion package with an explicit treatment of entrainment processes," Monthly Weather Review, vol. 134, no. 9, pp. 2318–2341, 2006.

21. J. E. Pleim, "A combined local and nonlocal closure model for the atmospheric boundary layer. Part II: application and evaluation in a mesoscale meteorological model," Journal of Applied Meteorology and Climatology, vol. 46, no. 9, pp. 1396–1409, 2007.

22. B. Storm, J. Dudhia, S. Basu, A. Swift, and I. Giammanco, "Evaluation of the weather research and forecasting model on forecasting low-level jets: implications for wind energy," Wind Energy, vol. 12, no. 1, pp. 81–90, 2009.

23. J. L. Woodward, "Appendix A: atmospheric stability classification schemes," in Estimating the Flammable Mass of a Vapor Cloud, pp. 209–212, American I. Wiley Online Library, 1998.

24. L. Fita, J. Fernández, M. García-Díez, and M. J. Gutiérrez, "SeaWind project: analysing the sensitivity on horizontal and vertical resolution on WRF simulations," in Proceedings of the 2nd Meeting on Meteorology and Climatology of the Western Mediterranean (JMCMO '10), 2010.

25. C. Talbot, E. Bou-Zeid, and J. Smith, "Nested mesoscale large-Eddy simulations with WRF: performance in real test cases," Journal of Hydrometeorology, vol. 13, no. 5, pp. 1421–1441, 2012.

26. S. Shimada, T. Ohsawa, T. Kobayashi, G. Steinfeld, J. Tambke, and D. Heinemann, "Comparison of offshore wind speed profiles simulated by six PBL schemes in the WRF model," in Proceedings of the 1st International Conference Energy & Meteorology Weather & Climate for the Energy Industry (IECM '11), pp. 1–11, November 2011.

27. K. E. Taylor, "Summarizing multiple aspects of model performance in a single diagram," Journal of Geophysical Research D, vol. 106, no. 7, pp. 7183–7192, 2001.

CHAPTER 5

A Subgrid Parameterization for Wind Turbines in Weather Prediction Models with an Application to Wind Resource Limits

B. H. FIEDLER AND A. S. ADAMS

5.1 INTRODUCTION

Wind power production in numerical weather prediction models can be either inert or active. In the inert type, the wind speed forecasted for a turbine location can be extracted from the model and used to calculate wind power production, with no impact of the turbines on the weather prediction [1]. In the active type, this impact is included, specifically the drag and turbulence enhancement of the wind turbine acting on the atmosphere [2]. In this paper we offer some details of a wind turbine parameterization appropriate for large wind farms, with many turbines within a grid cell. This paper refines the wind turbine parameterization in [2, 3], effectively offering a simplified and documented alternative to what appeared in WRFv3.3

A Subgrid Parameterization for Wind Turbines in Weather Prediction Models with an Application to Wind Resource Limits. © Fiedler BH and Adams AS. Advances in Meteorology **2014** (2014). http://dx.doi.org/10.1155/2014/696202. Licensed under a Creative Commons Attribution 3.0 Unported License, http://creativecommons.org/licenses/by/3.0/.

[4, 5]. Being subgrid, wakes are not explicitly simulated, but rather the momentum loss is immediately diffused across the breadth of the grid cell. The parameterization is adaptable to typical wind turbine characteristics. The giant wind farm of [3] is revisited for the purpose of studying the practical limit to wind power extraction from the atmosphere. Whereas [3] examined the much more subtle effect of the wind farm on precipitation climate statistics, the current study is more straightforward and does not require multidecadal simulations. The simulations use WRFv3.1 with the MYJ boundary layer scheme and 30 km horizontal grid spacing. The wind turbine parameterization adds elevated drag and production of turbulent kinetic energy to the MYJ scheme.

If a horizontal wind vector V is known at the height of wind turbine (in practice, meaning that a suitable average wind vector is known), then the power produced is

$$P = C_f(V) P_{max} \tag{1}$$

where C_f is the capacity coefficient and P_{max} is the rated power output for the particular wind turbine. In the simulations, we focus on $P_{max} = 2MW$ and $P_{max} = 8MW$, which roughly brackets the range of potential installations. C_f is constrained by both laws of nature and engineering design. For $V < V_{in}$, the turbine blades do not rotate, so $C_f = 0$. Likewise, for $V > V_{out}$ the turbine rotation is halted to avoid damage, and $C_f = 0$. As V increases past V_{in}, power production rises rapidly, but by engineering design is brought to a broad plateau of P_{max}, by mechanical adjustment of the turbine blade pitch angle [6].

Figure 1 shows a typical $C_f(V)$ offered by this parameterization. Figure 1 also shows the two other dimensionless coefficients that must be known, if the impact of the turbine on the atmosphere is to be calculated. The aerodynamical basis of (1) determines those impacts. From elementary physics, the available power of wind impinging on the rotor cross-sectional area A of the turbine is

$$\frac{1}{2}\rho V^3 A \qquad (2)$$

where ρ is the density of the air. The ratio of P to the available power is the power coefficient C_p:

$$C_p = \frac{C_f P_{max}}{\left(\frac{1}{2}\right)\rho V^3 A} \qquad (3)$$

So (1) can be written as

$$P = C_p(V)\frac{1}{2}\rho V^3 A \qquad (4)$$

Though (4) provides the identical calculation of power production as (1), knowledge of C_p will have another important purpose in calculating the production of turbulent kinetic energy.

The drag force F on an object presenting cross-sectional area A to a uniform stream of fluid with velocity V is conventionally modeled in terms of a shape-dependent drag coefficient C_d. In the language of wind turbine modeling, the drag coefficient is named the thrust coefficient C_t:

$$\vec{F} = C_t(V)\frac{1}{2}\rho V \vec{V} A \qquad (5)$$

At large Reynolds number, C_t is predominantly shape dependent. For example, there are many references giving values such as for a flat plate $C_t = 1.28$ and for a sphere $C_t = 0.47$. The drag force for rotating turbine blades is much greater than a calculation based on stationary blades and using just the area presented by the blades. The drag force of the wind turbine is characterized in terms of the disk swept out, $A = \pi R^2$, where R is blade length. For the Bonus Energy A/S 2.0 MW, C_t peaks at approximately $C_t = 0.88$ [2]. Presumably this cited value of C_t includes the drag of the tower as well, but in this parameterization the drag force is modeled as occurring within the area of the rotor.

The drag force of the wind turbine removes momentum from the atmosphere and transfers it to the Earth. But with the Earth having a large mass, the drag force transmitted to Earth, via the tower, does not do significant work on the Earth, meaning that the loss of energy from the mean wind goes into power production and turbulent kinetic energy, rather than into kinetic energy of the Earth [4].

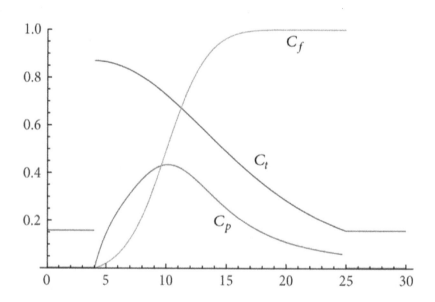

FIGURE 1: The capacity factor C_f, the thrust coefficient C_t, and power coefficient C_p for the wind turbine parameterization configured to model the Bonus Energy A/S 2 MW wind turbine. $V_{in} = 4$, $V_{out} = 25$, $\alpha = 0.3$, $\beta = 1.18 \times 10^{-5}$, $V_0 = 10$, $C_{ts} = 0.158$, and $C_{tp} = 0.87$.

The force of the turbine on the atmosphere is opposite to that of (5), so the rate of work (power) P_a on the atmosphere is $-F \times V$:

$$P_a = -C_t \frac{1}{2} \rho V^3 A \qquad (6)$$

By our principle of strict energy conservation

$$P + P_a + P_{tke} = 0 \qquad (7)$$

so

$$P_{tke} = \frac{1}{2} \rho V^3 A \left(C_t - C_p \right) \qquad (8)$$

Most numerical weather prediction models employ a force per mass at a grid point, within a grid volume. The drag force in (5) would need to be normalized appropriately, by the total mass of air in the grid volume. Similarly, a normalization is required when (8) is used to predict turbulent kinetic energy and added to the other source terms in the prediction for turbulent kinetic energy.

Here we take A as the only portion of the multiple wind turbine areas that are within the heights bounding a grid volume (Figure 2). This introduces some unrealism, as it allows a wind turbine to be modeled as having different rotation speeds and different C_f at various heights. The normalization procedure of the previous paragraph means that the rotor area per grid volume (an area density with units of inverse length) is the quantity needed for computation. In a staggered grid model, the heights of the prediction of horizontal wind may lie between the levels for the prediction of vertical velocity, the levels of which define the vertical bounds to the grid volume for horizontal velocity.

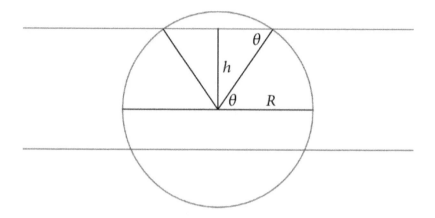

FIGURE 2: Calculating the fraction of the area of the rotor circle contained between two pressure levels. In this example, the hub of the rotor lies between the pressure levels. The area above the hub is sum of two triangles and two sectors. The area of the triangles sum to h $\sqrt{(R^2 - h^2)}$. The area of the two sectors sum to $R^2 \sin^{-1}(h/R)$. In this example, the area below the hub makes a similar positive contribution. If the hub is not between the layers, the total area is given by the subtraction of two areas calculated from the hub. Likewise, if $h > R$, then h is replaced by R in the calculation.

5.2 FUNCTIONS FOR $C_F(V)$ AND $C_T(V)$

For $C_f(V)$, we employ a soft-clip function, which allows $C_f(V)$ to come to a plateau without a sharp "knee point." Note that [4] does not provide this soft-clip feature and [2, 3] do not have a monotonic C_f. The soft-clip function that we employ is computationally efficient and provides a very close approximation to $(1 + \tanh(x))/2$:

$$s(x) = \begin{cases} 0 & \text{if } x \leq -3 \\ 1 & \text{if } x \geq 3 \\ \frac{1}{2}\left(1 + \dfrac{27x + x^3}{27 + 9x^2}\right) & \text{otherwise} \end{cases} \tag{9}$$

Here x = $\alpha(V - V_0)$, where V_0 controls the center point of C_f and α controls the slope of the transition. Adjusting the center point and slope to approximate the characteristics of the Bonus wind turbine is elementary. Slightly more complicated is to require C_p to be exactly zero for $V < V_{in}$, which may require employing a shift of s(x) to bury part of it below the x-axis. Let

$$\delta \equiv s[\alpha(V_{in} - V_0)] \tag{10}$$

Thus

$$C_f(V) = \begin{cases} 0 & \text{if } V \leq V_{in} \\ 0 & \text{if } V \geq V_{out} \\ \dfrac{1}{1-\delta}\{s[\alpha(V - V_0)] - \delta\} & \text{otherwise} \end{cases} \tag{11}$$

As the blades of a wind turbine are adjusted, to reduce C_p so that the maximum in C_f does not exceed 1, the thrust coefficient is also reduced. We find the following fit satisfactory:

$$C_t(V) = \begin{cases} C_{ts} & \text{if } V \leq V_{in} \\ C_{ts} & \text{if } V \geq V_{out} \\ C_{tp}\dfrac{1}{1 + .005(V - V_{in})^2 + \beta(V - V_{in})^4} & \text{otherwise} \end{cases} \tag{12}$$

Figure 1 lists the values of the parameters used for the Bonus 2 MW turbine. We employ a value for β that makes $C_t(V_{out}) = C_{ts}$, so there is no discontinuity in C_t at C_{out}. Other choices are possible.

5.3 APPLICATION TO WIND RESOURCE LIMITS

Textbooks in atmospheric science cite the typical midlatitude pressure gradient force (per mass) to be 10^{-3} m s^{-2} and the typical horizontal velocity scale to be 10 m s^{-1}, with about 1/10 of that being cross-isobaric. Only the cross-isobaric component is capable of internally renewing kinetic energy that has been removed by the wind farm. In the boundary layer, a larger fraction of the wind vector could be cross-isobaric, but the magnitude of the vector could be less. So if we accept 1 m s^{-1} as the typical magnitude of cross-isobaric flow, the rate of kinetic energy production within the wind farm would be 1 W m^{-2} per kilometer of depth of the extraction (assuming a density of $\rho = 1$ kg m^{-3}).

Kinetic energy also enters the wind farm from the side. If a giant wind farm of horizontal area LxL is extracting wind from a layer of depth H (where H could be the depth of the atmospheric boundary layer, rather than height to the top of the wind turbine), then (2) can be used to calculate the power advected into the wind farm presenting a side area of A = H x L. This is power that can be potentially extracted over the area of the wind farm, giving a power density

$$\frac{1}{2}\rho V^3 \frac{HL}{L^2} = \frac{1}{2}\rho V^3 \frac{H}{L} \tag{13}$$

For example, let us take H = 1 km and L = 100 km. For H/L = 0.01 and V = 5 m s^{-1}, that gives an upper limit to extraction of 0.625 W m^{-2}. Using V = 10 m s^{-1} instead gives 5 W m^{-2}. Note that a giant wind farm could have H/L < 0.01, and the bound on power that is extractable from the advection source would correspondingly be less. Thus as H/L becomes small, renewing of the wind resource by the pressure gradient becomes more important.

In the above estimates, what value should be used for H? Also, what is the contribution of transport through the top of the wind farm? We also need to recognize that as the power extraction approaches the upper limit, that would imply that V would be decreasing as the wind farm is traversed. The continuity equation would thus require upward advection of energy out of the top of the wind farm. All these considerations imply that a more refined estimate of the limits to power extraction at a site will require details about the wind climate, including boundary layer mixing, as well as the use of a numerical weather prediction model.

Here we demonstrate an application of our wind farm parameterization with a modest extension to several such studies of limits to wind farm resources [2, 4, 7], namely, allowing for all sizes of wind turbines to have the characteristics of Figure 1. The parameterization is used to investigate the relative effect of deploying the capacity density as 8 MW turbines, twice the size of 2 MW turbines.

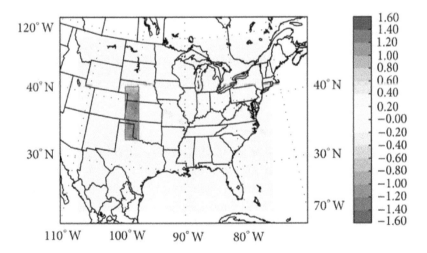

FIGURE 3: The wind farm is within the shaded rectangle. The average wind difference at 102 m between the simulation with 2.5 W m^{-2} of 2 MW turbines minus the simulation without turbines is shown.

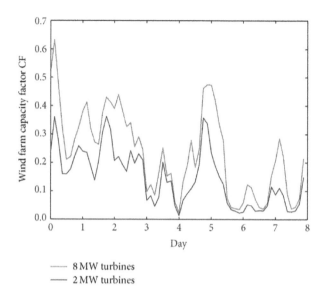

FIGURE 4: The average wind farm capacity factor CF for two particular deployments of CD = 2.5 W m^{-2}, as a function of time. Theses plots show the details behind the points for CD = 2.5 W m^{-2} in Figures 7 and 8. The average production with 8 MW turbines is 106 GW. The average production with 2 MW turbines is 66 GW.

The wind farm location is as in [3], with an area of 182,700 km^2 (Figure 3). In [3], 2 MW wind turbines were situated with turbine density of 1.25 km^{-2}, giving the giant wind farm a capacity of 457 GW and a capacity density of 2.5 W m^{-2}. The 2 MW wind turbines had a hub height of 60 m and a rotor radius of 38 m. Here we experiment with both 2 MW and 8 MW turbines, deployed with capacity density ranging from 0.625 W m^{-2} to 20 W m^{-2}. The 8 MW wind turbines are simply double in height and radius of the 2 MW turbines. The C_f, C_t, and C_p curves for both models are as in Figure 1. The study shown here is much simpler than [3], and examines only the effect of the wind farm on wind characteristics within the wind farm, as well as the power production by the wind farm.

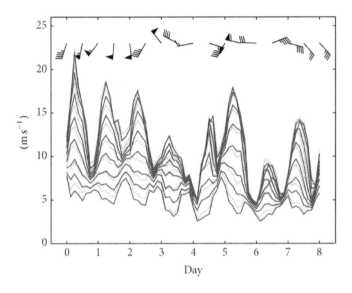

Figure 5: Average horizontal wind speed at the lowest 8 grid wind levels of 15, 52, 102, 165, 245, 345, 466, and 648 meters above ground. Lighter gray is without wind turbines. Darker color is with 2 MW wind turbines deployed at CD = 2.5 W m⁻² capacity. Top of the rotor is at 98 m (careful counting shows that darker curve is below the corresponding ligher curve). The wind barbs represent the horizontal wind direction and magnitude at 102 m at the center of the wind farm area, but without wind turbines. Half barbs are 1 m s⁻¹, full barbs 2 m s⁻¹, and flags 10 m s⁻¹. Days are ticked at 0 UTC, late afternoon at the wind farm, at which time the wind speed has become more well mixed across the boundary layer.

The 8 days from 0 UTC April 23, 1948 to 0 UTC May 1, 1948 were convenient for this study. The National Renewable Energy Laboratory displays the annual average wind speed at 80 m to range between 7.0 m s⁻¹ and 9.0 m s⁻¹ in the modeled wind farm area. The model, without wind turbines, has an average wind speed of 7.0 m s⁻¹ and 8.1 m s⁻¹ at 52 m and 102 m, respectively, during the 8 days of the simulation.

5.3.1 8 MW VERSUS 2 MW DEPLOYMENT

Here we highlight a particular comparison between deploying the 457 GW as either 228,375 2 MW turbines or as 58,656 8 MW turbines. In the analysis of power production, the average C_f experienced over the entire farm is denoted by CF (C_f is an engineering design parameter and CF is an experimental result). Though the capacity density (CD) is the same, the production density (PD) with the 8 MW deployment is 60% greater (Figure 4). A naive estimate might have anticipated an increase greater than 100%, using reasoning that the layer being mined for power is twice as deep, with the upper part having stronger winds. That sort of estimate is not realized.

Figures 5 and 6 show that the extraction (as indicated by wind speed reduction) has become rather insignificant at height 648 m. But both the 8 MW and 2 MW turbines discernibly remove energy below 648 m. As expected, the taller 8 MW turbines extract more energy in the layer above the height of the smaller turbine. Estimating how much more effective the extraction is with taller turbines has required the benefit of a numerical simulation. The ability to make such an estimation is one of the main practical benefits of the parameterization.

5.3.2 PRODUCTION SATURATION

Here we summarize the investigation into power production across a broad range of wind farm characteristics. Conclusions are similar to [2, 4, 7, 8]: Figure 7 shows a limit to power extraction to be on the order of 1 W m^{-2}. Such knowledge obviously influences design characteristics of wind farms: whether to add more wind turbines to a farm, acquire more land and develop a larger farm, or develop another farm in a distant location. In our simulations, the drop in CF proceeds immediately from the lowest CD. This is because power production is very sensitive to changes in V in the vicinity of V_{in}, with C_f rising faster than V^3 wherever C_p is increasing with V. The opposite scenario could happen in a different wind climate. If V is consistently well into the range that produces $C_f(V) = 1$, there might be a significant decrease in V, but no drop in CF until CD exceeds a value greater than 1 W m^{-2}.

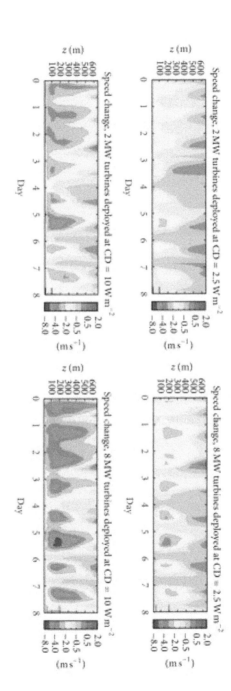

FIGURE 6: Area-averaged wind speed change that results from deploying wind turbines, as a function of height and time. The height of the top and bottom of the rotor are indicated by longer tick marks at the extreme right. The extent of the wind reduction above the tops of the turbines is indicative of power extraction from those layers, a complicated prediction requiring a numerical weather prediction model.

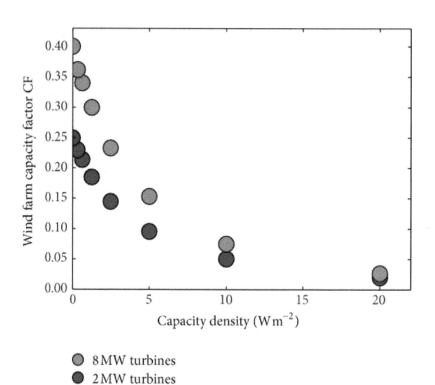

FIGURE 7: As in Figure 8, but average wind farm capacity factor CF for various deployments.

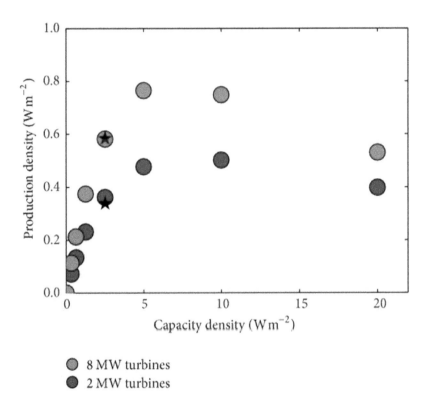

FIGURE 8: Average wind farm production density PD for various deployments. The two points marked with a star are for simulations repeated using twice the number of grid points within the depth of the boundary layer, as compared with the standard resolution. The simulations with standard resolution are indicated with a circle. The vertical positions of the grid points in the standard resolution are listed in Figure 5.

Consider increasing CD from 2.5 W m^{-2}, with 2 MW turbines, to 10 W m^{-2} by either quadrupling the area of the rotors or quadrupling the number of turbines. The two scenarios can be found within Figure 8. Increasing the rotor area density by a factor of 4 (redeploying as 8 MW turbines) increased PD by 2.07. Quadrupling the number of 2 MW turbines increased PD by a factor of 1.38. Since the increase in PD was significantly less than 4, we would say that the collective impact of the turbines on the power productivity of the winds is significant. Inspection of the wind difference plots in Figure 6 shows wind being reduced above the tops of the turbines, evidently the effect of turbulent transport of momentum vertically in the atmosphere. This transport may be hard to estimate by means other than a detailed numerical model.

We note that Horns Rev 1, an established 20 km^2 wind farm in the North Sea, has been averaging PD = 3.98 W m^{-2} in the last 5 years [9]. This illustrates the importance of scale in understanding the production limitations of wind energy. As discussed in Section 3, the larger the horizontal extent of the wind farm is, the less important the advection of kinetic energy is and the more important the pressure gradient force becomes in sustaining energy producing wind speeds within the wind farm. The Horns Rev 1 wind farm is small enough that H/L is approximately 0.2, thus explaining the observed PD.

5.4 CONCLUSIONS

When considering national and international energy portfolios, wind energy continues to become an important part of diversified energy portfolios. Though current wind farms are small enough in scale to have H/L ratios that allow advection of kinetic energy into the side of a wind farm to be an important power source, it is important to discern how that wind resource diminishes with larger wind farms. Future power needs could force the development of giant wind farms, with areas that are orders of magnitude larger than current farms. Furthermore, the development of many small wind farms in close proximity could have a resource limit similar to a giant wind farm.

Giant wind farms will need to be planned with an active type numerical weather prediction model, so as to get an accurate estimate of the wind power resource. For example, in our study, larger (taller) wind turbines produce a larger CF. The cost-effectiveness of deploying larger turbines will require an accurate prediction of this CF before a financial decision can be made.

REFERENCES

1. J. S. Greene, M. Chatelain, M. Morrissey, and S. Stadler, "Projected future wind speed and wind power density trends over the western us high plains," Atmospheric and Climate Sciences, vol. 2, no. 1, pp. 32–40, 2012.
2. A. S. Adams and D. W. Keith, "Are global wind power resource estimates overstated?" Environmental Research Letters, vol. 8, no. 1, Article ID 015021, 2013.
3. B. H. Fiedler and M. S. Bukovsky, "The effect of a giant wind farm on precipitation in a regional climate model," Environmental Research Letters, vol. 6, no. 4, Article ID 045101, 2011.
4. A. C. Fitch, J. B. Olson, J. K. Lundquist et al., "Local and mesoscale impacts of wind farms as parameterized in a mesoscale nwp model," Monthly Weather Review, vol. 140, no. 9, pp. 3017–3038, 2012.
5. W. C. Skamarock and A. Coauthors, "description of the advanced research WRF version 3," Tech. Rep. NCAR/TN-4751STR, National Center for Atmospheric Research, 2008.
6. P. W. Carlin, A. S. Laxson, and E. B. Muljadi, "The history and state of the art of variable-speed wind turbine technology," Wind Energy, vol. 6, no. 2, pp. 129–159, 2003.
7. D. W. Keith, J. F. DeCarolis, D. C. Denkenberger et al., "The influence of large-scale wind power on global climate," Proceedings of the National Academy of Sciences of the United States of America, vol. 101, no. 46, pp. 16115–16120, 2004.
8. M. Z. Jacobson and C. L. Archer, "Saturation wind power potential and its implications for wind energy," Proceedings of the National Academy of Sciences, vol. 109, no. 39, pp. 15679–15684, 2012.
9. E. L. Petersen, I. Troen, H. E. Jrgensen, and J. Mann, "Are local wind power resources well estimated?" Environmental Research Letters, vol. 8, no. 1, Article ID 011005, 2013.

CHAPTER 6

Are Local Wind Power Resources Well-Estimated?

ERIK LUNDTANG PETERSEN, IB TROEN, HANS E. JØRGENSEN, AND JAKOB MANN

Planning and financing of wind power installations require very importantly accurate resource estimation in addition to a number of other considerations relating to environment and economy. Furthermore, individual wind energy installations cannot in general be seen in isolation.

It is well known that the spacing of turbines in wind farms is critical for maximum power production. It is also well established that the collective effect of wind turbines in large wind farms or of several wind farms can limit the wind power extraction downwind. This has been documented by many years of production statistics. For the very large, regional sized wind farms, a number of numerical studies have pointed to additional adverse changes to the regional wind climate, most recently by the detailed studies of Adams and Keith [1]. They show that the geophysical limit to wind power production is likely to be lower than previously estimated.

Are Local Wind Power Resources Well Estimated?. © *Petersen EL, Troen I, Jørgensen HE, and Mann J*. Environmental Research Letters *8 (2013), doi:10.1088/1748-9326/8/1/011005. Licensed under Creative Commons Attribution 3.0 Unported License, http://creativecommons.org/licenses/by/3.0/.*

Although this problem is of far future concern, it has to be considered seriously. In their paper they estimate that a wind farm larger than 100 km^2 is limited to about 1 W m^{-2} . However, a 20 km^2 off shore farm, Horns Rev 1, has in the last five years produced 3.98 W m^{-2} [5]. In that light it is highly unlikely that the effects pointed out by [1] will pose any immediate threat to wind energy in coming decades.

Today a number of well-established mesoscale and microscale models exist for estimating wind resources and design parameters and in many cases they work well. This is especially true if good local data are available for calibrating the models or for their validation.

The wind energy industry is still troubled by many projects showing considerable negative discrepancies between calculated and actually experienced production numbers and operating conditions. Therefore it has been decided on a European Union level to launch a project, 'The New European Wind Atlas', aiming at reducing overall uncertainties in determining wind conditions.

The project is structured around three areas of work, to be implemented in parallel.

- Creation and publication of a European wind atlas in electronic form [2], which will include the underlying data and a new EU wind climate database which will as a minimum include: wind resources and their associated uncertainty; extreme wind and uncertainty; turbulence characteristics; adverse weather conditions such as heavy icing, electrical storms and so on together with the probability of occurrence; the level of predictability for short-term forecasting and assessment of uncertainties; guidelines and best practices for the use of data especially for micro-siting.
- Development of dynamical downscaling methodologies and open-source models validated through measurement campaigns, to enable the provision of accurate wind resource and external wind load climatology and short-term prediction at high spatial resolution and covering Europe. The developed downscaling methodologies and models will be fully documented and made publicly available and will be used to produce overview maps of wind resources and other relevant data at several heights and at high horizontal resolution.
- Measurement campaigns to validate the model chain used in the wind atlas. At least five coordinated measurement campaigns will be undertaken and will cover complex terrains (mountains and forests), offshore, large changes in surface characteristics (roughness change) and cold climates.

One of the great challenges to the project is the application of mesoscale models for wind resource calculation, which is by no means a simple matter [3]. The project will use global reanalysis data as boundary conditions. These datasets, which are time series of the large-scale meteorological situation covering decades, have been created by assimilation of measurement data from around the globe in a dynamical consistent fashion using large-scale numerical models. For wind energy, the application of the reanalysis datasets is as a long record of the large-scale wind conditions. The large-scale reanalyses are performed in only a few global weather prediction centres using models that have been developed over many years, and which are still being developed and validated and are being used in operational services. Mesoscale models are more diverse, but nowadays quite a number have a proven track record in applications such as regional weather prediction and also wind resource assessment. There are still some issues, and use of model results without proper validation may lead to gross errors. For resource assessment it is necessary to include direct validation with in situ observed wind data over sufficiently long periods. In doing so, however, the mesoscale model output must be downscaled using some microscale physical or empirical/statistical model. That downscaling process is not straightforward, and the microscale models themselves tend to disagree in some terrain types as shown by recent blind tests [4]. All these 'technical' details and choices, not to mention the model formulation itself, the numerical schemes used, and the effective spatial and temporal resolution, can have a significant impact on the results. These problems, as well as the problem of how uncertainties are propagated through the model chain to the calculated wind resources, are central in the work with the New European Wind Atlas. The work of [1] shows that when wind energy has been implemented on a very massive scale, it will affect the power production from entire regions and that has to be taken into account.

REFERENCES

1. Adams A S and Keith D W 2013 Are global wind power resource estimates overstated? Environ. Res. Lett. 8 015021

2.   2011 A New EU Wind Energy Atlas: Proposal for an ERANET+ Project (Produced by the TPWind Secretariat) Nov.

3.   Petersen E L and Troen I 2012 Wind conditions and resource assessment WIREs Energy Environ. 1 206–17

4.   Bechmann A, Sørensen N N, Berg J, Mann J and Rethore P-E 2011 The Bolund experiment, part II: blind comparison of microscale flow models Boundary-Layer Meteorol. 141 245–71

5.   www.lorc.dk/offshore-wind-farms-map/horns-rev-1

6.   www.ens.dk

PART III

SOLAR ENERGY AND THE WEATHER

CHAPTER 7

Impact of Terrain and Cloud Cover on the Distribution of Incoming Direct Solar Radiation over Pakistan

SHAHZAD SULTAN, RENGUANG WU, AND IFTIKHAR AHMED

7.1 INTRODUCTION

Distributing modeling is the emerging technique to analyze the spatial and temporal variability of solar radiation, particularly over the rugged terrain. The knowledge of the solar radiation distribution over time and space is becoming more and more vital due to the growing interest in solar energy utilization [1]. Incoming direct solar radiation (DSR) is the main contributor of incoming solar radiation to earth because it adds the most to the energy balance and also the other components depend on it [2]. The understanding of the DSR distribution both temporally and spatially is one of the imperative needs for understanding of thermody-

Impact of Terrain and Cloud Cover on the Distribution of Incoming Direct Solar Radiation over Pakistan. © Sultan S, Wu R, and Iftikhar Ahmed I. Journal of Geographic Information System *6,1 (2014), OI:10.4236/jgis.2014.61008. Licensed under Creative Commons Attribution License, http://creativecommons.org/licenses/by/3.0/.*

namic or dynamic processes involved in energy exchange, which can be applicable in agricultural planning [3], architectural design [4], and engineering [5]. Solar radiation energy exchange also plays a key role in surface water circulation [6]. For these reasons, analysis of the incoming solar radiation distribution in Pakistan is important. As seasonal and latitudinal variations are well understood and described; only recent studies analyzed topographic variability in diurnal and seasonal radiation patterns over a region or locally [7].

Topography is one of the major factors determining the amount of incident solar energy radiation at a particular location [8]. The complexity of topography due to variations in altitude, slope, aspect and surface nature makes the radiation spatial distribution more complicated [9]. Pakistan's north-south extension with complex and diversified terrain conditions and scarce network of observatories makes it difficult to measure the distribution of incoming solar radiation through an existing ground based network of observations.

So far, many domestic and international solar radiation estimation models based on flat surface observed data have been established [10]. There are many factors which make topographic solar radiation modeling more complex, such as complexity of physical based solar radiation formulation, insufficient observed data and lack of suitable modeling tools [8]. Since the late seventies of the last century, a series of distributed models for land surface have been developed rapidly, but the progress in the solar radiation estimation over the rugged terrain is not much significant [11-13]. Therefore, to explore new ways of estimating solar radiation over the undulating terrain and other space-time spatial and temporal distribution of volatile elements is of great theoretical significance and has wide application prospects. Till now, there is no published work on the distribution of DSR linked with terrain effect over Pakistan.

7.2 STUDY AREA

Pakistan lies between latitudes 24°N - 37°N and longitudes 62°E - 75°E, covering a total land area of 796,096 km^2 (www.gov.pk). The geography of Pakistan is a blend of landscapes varying from plains to deserts, forests, hills, and plateaus ranging from the coastal areas of the Arabian Sea in the south to the mountains of the HinduKush, Karakoram Himalaya range in the north. Pakistan is divided into three major geographic areas: the northern highlands; the Indus River plain, and the Baluchistan Plateau. The northern highlands of Pakistan contain the Karakoram, Hindu Kush and Pamir mountain ranges and some of the world's highest peaks, including K2 (8611 m /28,251 ft) and Nanga Parbat (8126 m/26,660 ft). The Baluchistan Plateau lies to the west, and the Thar Desert in the east. An expanse of alluvial plains lies in Punjab and Sindh along the Indus River. The 1609 km (1000 mi) Indus River and its tributaries flow through the country from the Kashmir region to the Arabian Sea [14,15] Furthermore, 23.8% area of Pakistan is semi-arid, 56% is arid, 3.6% is mesodermal with dry winter and 16.1% is alpine [16] (Figure 1).

7.3 DATA AND METHOD

7.3.1 METEOROLOGICAL DATA

About 22 years data of daily global incoming solar radiation (1978-2000), 29 years data of monthly cloud cover and monthly sunshine duration (1979-2008) are provided by Pakistan Meteorological Department. The data for cloud cover and sunshine duration were processed for 21 stations, while for solar radiation, 6 meteorological observatories data from Pakistan (only Global Solar Radiation) and 5 meteorological stations from China (Direct and Global Solar Radiation) were processed. Data from Pakistan remain the pronounced concern of apprehension throughout the research period in terms of quality and availability. The continuity the data of solar radiation can be categorized into two groups, one from 1979 to 2000 and other from 2001 to 2008. The data in group one is used for simulation due to relatively better consistency in terms of availability, while the data from 2001 to 2008 is used to compare the results (only for Global Radiation).

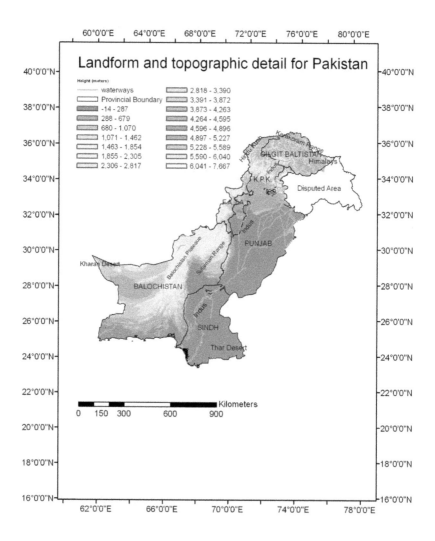

FIGURE 1: Landform details for Pakistan.

7.3.2 SATELLITE AND DEM DATA

DEM over Pakistan is obtained from SRTM (Shuttle Radar Topography Mission at the website ftp://e0mss21u.ecs.nasa.gov/srtm. The data of 3 ArcSecond, which was obtained almost with 90m cell size, is employed in this paper. Moderate Resolution Imaging Spectroradiometer (MODIS) Level-2 cloud fraction product data were used to estimate cloud cover over Pakistan at a resolution of about 900m × 900m.

7.4 METHODOLOGY

To incorporate the cloud cover effect on DSR, MODIS cloud fraction data were processed and rectified using ground observed data of cloud cover on monthly mean basis using ArcGIS. Also, possible sunshine duration was corrected by DEM and used with rectified MODIS cloud fraction data to calculate the linear regression for obtaining fitted empirical coefficients for direct solar radiation.

According to the model given by [17] based on the theory of direct solar radiation over inclined surface [18], we have the following relation,

$$\frac{H_{0\alpha\beta}}{H_0} = \frac{H_{b\alpha\beta}}{H_b} \tag{1}$$

where $H_{0\alpha\beta}$ is the extraterrestrial solar radiation (ESR) quantity on slope, $H_{b\alpha\beta}$ is incoming direct solar radiation (DSR) over rugged terrain, H_0 is ESR on flat surface and H_b is direct incident solar radiation on flat surface. Given the values of $H_{0\alpha\beta}$, H_0 and H_b, we get the distribution of direct solar radiation over rugged terrain.

Here, $H_{0\alpha\beta}$ is calculated by using the methodology given by [19], i.e.,

$$H_{0\alpha\beta} = \frac{T}{2\pi}\left(\frac{1}{\rho}\right)^2 I_0[u\sin\delta(\omega_{ss} - \omega_{sr}) + v\cos\delta(\sin\omega_{ss} - \sin\omega_{sr}) + w\cos\delta(\cos\omega_{ss} - \cos\omega_{sr})]$$

where $(1/\rho)^2$ is sun-earth distance correction factor, δ is solar declination, ω_{sr} is solar hour angle at the beginning time of the possible sunshine duration (PSD), ω_{ss} is solar hour angle at the ending time of the PSD, $H_{0\alpha\beta}$ is ESR quantity falling on the slope from ω_{sr} to ω_{ss}, and T is the total time length of one day, i.e. 1440 min. The physical meaning of this equation is that ESR on a slope is determined by geographic, topographic and astronomic factors. Using PSD as the range of integration $[\omega_{sr}, \omega_{s}]$.

H_0 is calculated by theoretical formulae [20], i.e.,

$$H_0 = \frac{L}{\pi}\left(\frac{1}{\rho}\right)^2 I_0 \cos\phi \cos\delta(\sin\omega_0 - \omega_0 \cos\omega_0)$$

where H_0 is ESR amount on horizontal surface, L is the time spac, I_0 is the solar constant and H_b is calculated by meteorological data fitting statistical models [17], the detail of which is given below.

7.4.1 HORIZONTAL DIRECT INCOMING RADIATION (H_B) MODEL

The model used for calculating direct solar radiation is given by [17].

$$H_b = H(1-a)\left(1 - \exp\left[\frac{1-bs^c}{(1-s)}\right]\right)$$
(2)

where H is global solar radiation quantity of flat surface, s is relative sunshine duration, a, b, and c are empirical coefficients.

DSR observational data in Pakistan is not sufficient. Therefore, the compensatory method must be employed. Monthly DSRs of 5 stations of China were used for calculating DSR in Pakistan. Those 5 stations are located in China, having the semi-arid climatic conditions.

In Table 1, the statistical indices are square of correlation coefficient R^2 and empirical coefficients a, b and c are given. During the study it has been found that monthly models (observational data of all stations of the same month are used) have higher accuracy and stability as compared to unified model (observational data of all stations are used to fit the model).

7.4.2 AVERAGE GLOBAL HORIZONTAL RADIATION (H) MODEL

Various studies show that global solar radiation (H) and relative sunshine duration (s) are closely related [21,22] by using Angstrom-Prescott equation:

$$H = H_0(a_c + b_c \cdot s)$$

(3)

where a_c and b_c are empirical coefficients.

Using the close-fitting empirical coefficients of the 5 meteorological stations from China as given in Table 1, we get S (relative sunshine duration) of remote model and H_0 horizontal global solar radiation simulation.

By using Equations (2) and (3), we get the following relation.

$$H_b = H_0(a_c + b_c \cdot s)(1 - a)\left(1 - \exp\left[\frac{-bs^c}{(1 - s)}\right]\right)$$

(4)

This equation only needs relative sunshine duration.

7.4.3. STATISTICS FOR DIFFERENT MODEL FOR GLOBAL SOLAR RADIATION

Table 2 shows the statistics of different models to get the optimized values for empirical coefficients a, and b to fit global radiation model (Equation (3)). Model-1 is the synthetic model considering 6 stations from Pakistan and 3 stations from China. Models 6-8 are offered by other scholars. Considering all these factors (e.g. fitting precision, sample amount, and relativity), Model-4 has the most optimal values as compared to others. Therefore, Model-4 values are considered to solve the average global horizontal radiation model (Equation (3)).

7.5 RESULTS AND DISCUSSION

By using Equation (1), we get the DSR for the years 2006 & 2007 (mean monthly values are considered here) with resolution of 900m × 900m grid cell by incorporating DEM to analyze the terrain effect (Figure 2). The increasing trend from north to south, that is, latitudinal dependency is common in all resultsThe minimum annual values of DSR lies in Gilgit Baltistan and the surrounding areas, ranging from 727 to 1687 MJ×m^{-2}; while the maximum annual values are located in deserted areas of Baluchistan, Sindh and Punjab provinces, ranging from 5036 to 5720 MJ×m^{-2}.

Figure 3 shows the distribution of DSR over the mountainous region of Pakistan in Indus upstream. There is a clear uneven distribution of DSR due to complex terrain, indicating the effect of local topography factors. The annual DSR quantity of sunny slope (or southern slope) of mountains is obviously larger than that of shady slope (or northern slope).

The observed data for DSR in Pakistan is only available for Quetta region. Therefore, results of simulation were being compared with high resolution solar maps produced by National Renewable Energy Laboratory [23] (Figure 4). The results with resolution of 4 km provided by NREL's 2010 show a consistent variations except for Quetta region (North west of Pakistan) for annula DSR. The model based on equation 2 having resolution of 900 m by 900 m is relatively underestimating the DSR as compared to the NREL's model. The possible reasons of less estimation of DSR are as follows. NREL's simulation resolution is coarser, about 4 km; NREL's simulation do not consider corrected cloud cover data; and NREL's simulation also do not consider corrected data of possible sunshine duration based on topography. This is also revealed by a comparison of actual DSR data of Quetta region with NREL's, which is overestimated by about 38%.

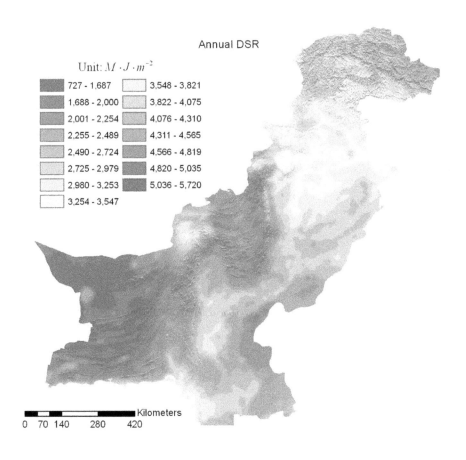

FIGURE 2: Annual distribution of DSR (mean of 2006 & 2007) over rugged terrain of Pakistan.

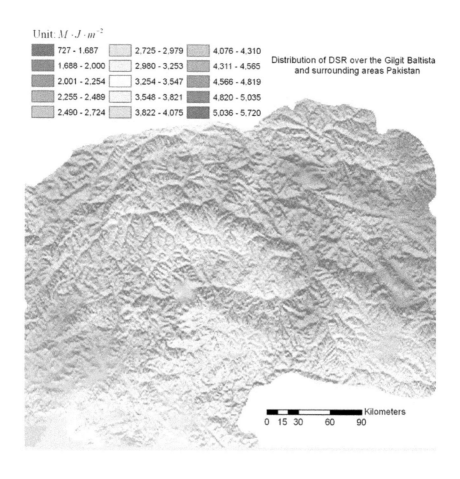

Unit: $M \cdot J \cdot m^{-2}$

727 - 1,687	2,725 - 2,979	4,076 - 4,310
1,688 - 2,000	2,980 - 3,253	4,311 - 4,565
2,001 - 2,254	3,254 - 3,547	4,566 - 4,819
2,255 - 2,489	3,548 - 3,821	4,820 - 5,035
2,490 - 2,724	3,822 - 4,075	5,036 - 5,720

Distribution of DSR over the Gilgit Baltista
and surrounding areas Pakistan

FIGURE 3: Distribution of direct solar radiation (mean of 2006 & 2007) over the mountainous region of Pakistan (Gilgit Baltistan).

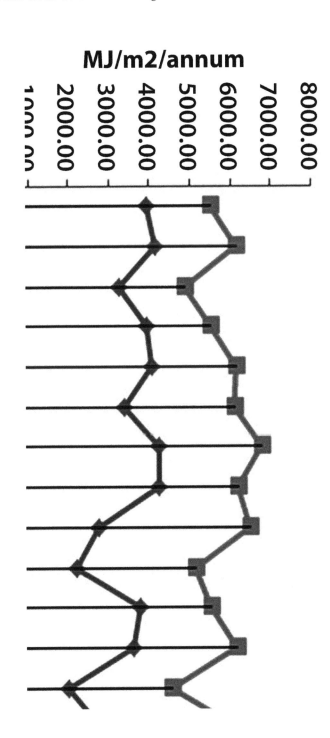

FIGURE 4: Comparison of simulated results of DSR using Equation (2) and USAID (NREL's, 2010).

Unit: $M \cdot J \cdot m^{-2}$

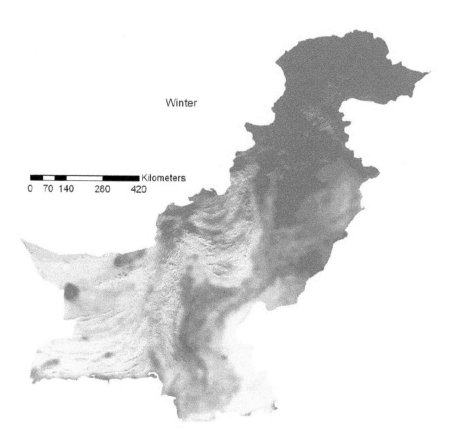

Winter (December, January, February)

FIGURE 5: Mean seasonal variations of direct solar radiation for the year of 2006 & 2007.

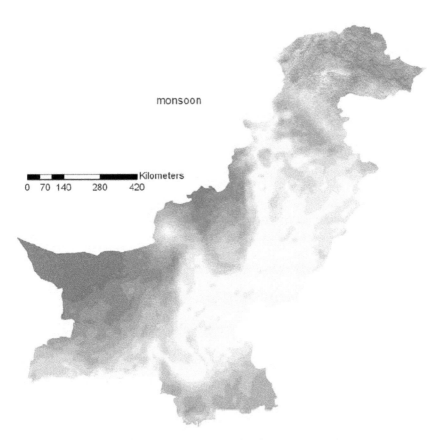

Monsoon (June, July, August)

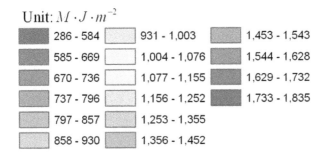

Unit: $M \cdot J \cdot m^{-2}$

286 - 584	931 - 1,003	1,453 - 1,543
585 - 669	1,004 - 1,076	1,544 - 1,628
670 - 736	1,077 - 1,155	1,629 - 1,732
737 - 796	1,156 - 1,252	1,733 - 1,835
797 - 857	1,253 - 1,355	
858 - 930	1,356 - 1,452	

FIGURE 5: *Cont.*

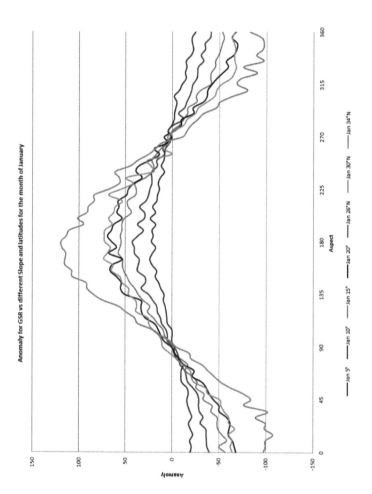

FIGURE 6: Topography effect over direct solar radiation at different slopes and latitudes for the month of January (anomaly is in MJ×m^{-2}).

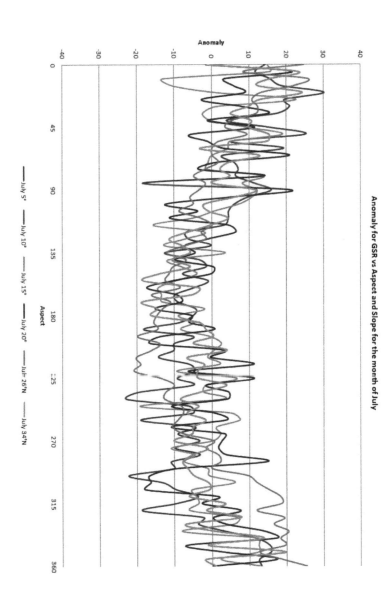

FIGURE 7: Topography effects over direct solar radiation on different slopes for the month of July (anomaly is in MJ×m⁻²). Natural Science Foundation of China grants (41228006 and 41275081).

7.5.1 SEASONAL VARIATIONS

Figure 5 depicts the mean seasonal variations of DSR for the year 2006 and 2007 over the rugged terrain of Pakistan. We compare the quantity of DSR during the monsoon and winter seasons. It is evident from the result that the quantity of DSR during monsoon season is higher than that during winter season. The southwestern part of Pakistan is least influenced by the monsoon and has higher values of DSR as compared to eastern parts of Pakistan, ranging from 1733 to 1835 MJ×m^{-2}. It is also obvious that the southeastern Sindh of Pakistan has comparatively lower values of DSR during summer than its surrounding areas. One of the reasons is cloud cover due to monsoonal flows and the other reason might be aerosols and dust particles [24].

7.5.2 TOPOGRAPHIC INFLUENCES ON DSR OVER THE RUGGED TERRAIN OF PAKISTAN

Figures 6 and 7 show the variations of DSR over different slopes (5, 10, 15 and 20 degree) and latitudes (26°N, 30°N and 34°N) at different azimuth (denotes east azimuth, south and west azimuth) for the months of January and July respectively. It has been observed that the variations during January represent the average trend for winter and those during July represent the average variation during summer. During January the effect of aspects over DSR leads to much larger fluctations as compared to July (Figures 6 and 7). Therefore, we can conclude that during winter southern slopes receive more DSR as compared to summer due to sun orientation. For eastern and western aspects, the variations of DSI are not so prompt, both in summer and winter, Therefore, eastern and western slopes receive almost the same amount of DSR throughout the year.

7.6 CONCLUSION

There is a significant impact of topography on the distribution of DSR over the northern areas of Pakistan, particularly in Gilgit Baltistan, Potwar plateau and Blochistan plateau. Overall, southern slopes receive more DSR as compared to horizontal surface during winter, and vice versa during summer. Topographic effects for eastern and western slopes are not significant and they receive almost the same solar irradiance due to the orientation of topography with respect to sun position and latitudnal consistency. During the monsoon season, DSR displays a tendencey of decrease from western side of the Indus River to eastern side of the Indus River due to distribution of cloud cover. Therefore, any estimation of solar energy or estimation of evapotranpiration in northern areas of Pakistan should also consider the topographic effects (slope, aspect, etc.).

REFERENCES

1. R. Chant and D. Ruth, "Solar Heating in Canada, The Potential of Solar Energy for Canada," Conference of the Solar Energy Society of Canada Inc, Ottawa, June 1975. http://www.cmos.ca/CB/cb100201.pdf
2. L. Wang and X. Qiu, "Distributed Modeling of Direct Solar Radiation of Rugged Terrain Based on GIS," The 1st International Conference on Information Science and Engineering (ICISE), Nanjing, 26-28 December 2009, 2042-2045.
3. S. Changnon and D. Changnon, "Importance of Sky Conditions on the Record 2004 Midwestern Crop Yields," Physical Geography, Vol. 26, No. 2, 2005, pp. 99-111.
4. Z. X. L. Yang and Y. F. Hu, "Study on Solar Radiation and Energy Efficiency of Building Glass System," Applied Thermal Engineering, Vol. 26, No. 8-9, 2006, pp. 956-961.
5. E. Amer and M. Younes, "Estimating the Monthly Discharge of a Photovoltaic Water Pumping System: Model Verification," Energy Conversion and Management, Vol. 47, No. 15-16, 2006, pp. 2092-2102.
6. R. Ranzi and R. Rosso, "A Stokesian Model of Areal Clear-Sky Direct Radiation for Mountainous Terrain," Geographical Research Letters, Vol. 20, No. 24, 1993, pp. 2893-2896. http://dx.doi.org/10.1029/93GL03307
7. Q. S. I. Dozier, "An Approach toward Energy Balance Simulation over Rugged Terrain," Geography, Vol. 11, No. 1, 1979, pp. 65-85.
8. R. Dubayaha and P. M. Richb, "Topographic Solar Radiation Models for GIS," International Journal of Geographical Information Science, Vol. 9, No. 4, 1995, pp. 405-419.
9. D. L. Liu and B. J. Scott, "Estimation of Solar Radiation in Australia from Rainfall and Temperature Observations," Agricultural and Forest Meteorology, Vol. 106, No. 1, 2001, pp. 41-59.
10. H. Yoo, K. Lee, S. Park and K. H. Noh, "Calculation of Global Solar Radiation Based on Cloud Data for Major Cities of South Korea," Global Warming, Green Energy and Technology, 2010, pp. 467-484.

11. R. Dubayah, "Estimating Net Solar Radiation Using Landsat Thematic Mapper and Digital Elevation Data," Water Resources Research, Vol. 28, No. 9, 1992, pp. 2469-2484. http://dx.doi.org/10.1029/92WR00772

12. R. Dubayah, J. Dozier and F. Davis, "The Distribution of Clear-Sky Radiation over Varying Terrain," 12th Canadian Symposium on Remote Sensing (IGARSS'89), Vancouver, 10-14 July 1989, pp. 885-888.

13. W. Panshou, "Mountain Climate Research Overview: On the Study of the Progress of the Mountain Climate, Mountain Climate Collected Works," Beijing Meteorological Press, Beijing, 1984, pp. 1-7.

14. P. Blood, "Pakistan: A Country Study," Federal Research Division of the Library of Congress, 1994. http://countrystudies.us/pakistan/23.htm

15. N. K. Amir, "Climate Change Adaptation and Disaster Risk Reduction in Pakistan," In: J. M. P. J. J. P. Rajib Shaw, Ed., Climate Change Adaptation and Disaster Risk Reduction: An Asian Perspective, Vol. 5, Emerald Group Publishing Limited, 2010, pp. 197-215.

16. climatemps.com, "Climate, Average Weather of Pakistan," 2013. http://www.pakistan.climatemps.com/

17. Y. Zeng, X. Qiu, C. Liu and A. Jiang, "Distributed Modeling of Direct Solar Radiation on Rugged Terrain of the Yellow River Basin," Journal of Geographical Sciences, Vol. 15, No. 4, 2005, pp. 439-447. http://dx.doi.org/10.1007/BF02892151

18. B. Liu and R. Jordan, "The Interrelationship and Characteristic Distribution of Direct, Diffuse and Total Solar Radiation," Solar Energy, Vol. 4, No. 3, 1960, pp. 1-19. http://dx.doi.org/10.1016/0038-092X(60)90062-1

19. X. Qiu, Y. Zeng and C. Liu, "Simulation of Astronomical Solar Radiation over Yellow River Basin Based on DEM," Journal of Geographical Sciences, Vol. 14, No. 1, 2004, pp. 63-69. http://dx.doi.org/10.1007/BF02873092

20. D. Zuo, Y. Zhou and Y. Xiang, "On Surface Radiations," Science Press, Beijing, 1991. (in Chinese)

21. L. T. Wong and W. Chow, "Solar Radiation Model," Applied Energy, Vol. 69, No. 3, 2001, pp. 191-224. http://dx.doi.org/10.1016/S0306-2619(01)00012-5

22. D. Weng, "Studies on Radiation Climate of China," China Meteorological Press, Beijing, 1997. (in Chinese)

23. NREL's, "Pakistan Resource Maps and Toolkit," 2010. http://www.nrel.gov/international/ra_pakistan.html?print

24. M. Rashed and L. Shuanglin, "Response of Summer Rainfall in Pakistan to Dust Aerosols in an Atmospheric General Circulation Model," Quarterly Journal of Hungarian Meteorological Science, Vol. 116, No. 4, 2012, pp. 323-333.

There are two tables that are not available in this version of the article. To view this additional information, please use the citation on the first page of this chapter.

CHAPTER 8

Trends in Downward Solar Radiation at the Surface over North America from Climate Model Projections and Implications for Solar Energy

GERARDO ANDRES SAENZ AND HUEI-PING HUANG

8.1 INTRODUCTION

The last decade has witnessed a rapid development in solar energy as an alternative to fossil-fuel based energy. Solar power plants with increasing size and efficiency have been built. With an increased stake in the investment and return, site selection and assessments of long-term sustainability for solar power plants become increasingly important. One of the factors that affect the long-term planning for solar energy is the local climatology. Long hours of sunshine at a location are essential for a viable solar power plant. The available solar energy at a given site is quantified by the downward solar (shortwave) radiation at the surface. At a given latitude and day of the year, this quantity is affected by atmospheric water vapor

Trends in Downward Solar Radiation at the Surface over North America from Climate Model Projections and Implications for Solar Energy. © *Saenz GA and Huang H-P.* Advances in Meteorology **2015** *http://dx.doi.org/10.1155/2015/483679. Licensed under Creative Commons Attribution 3.0 Unported License, http://creativecommons.org/licenses/by/3.0/.*

and trace gases, the amount of aerosols in the atmosphere, and, most importantly, cloud cover (e.g., Li et al. [1]). Considering those factors, climatological maps of downward solar radiation have been widely produced for solar energy applications (e.g., National Renewable Energy Laboratory, http://www.nrel.gov/gis/solar.html, Maxwell et al. [2], and George and Maxwell [3]). Since climate is constantly changing due to anthropogenic and natural processes, the estimates of solar power potential based on present-day climatology are not guaranteed to be true in the future. In this study, we will analyze the projection of the changes in the downward solar radiation in the 21st century over North America using a set of climate model simulations driven by anthropogenic greenhouse-gas (GHG) forcing from the Climate Model Intercomparison Project—Phase 3 (CMIP3) archive (Meehl et al. [4]). The global climate models have relatively coarse horizontal resolutions but are capable of producing the first-order features of atmospheric general circulation. Over North America, GHG-induced changes in the large-scale circulation are known to produce future drying in the Southwest USA and a poleward shift of storm tracks over Western USA (e.g., Seager et al. [5] and Baker and Huang [6]). These changes potentially imply more sunshine in the Southwest USA but reduced sunshine in the higher latitudes in Western USA due to increased cloudiness associated with storms. We will quantify the extent to which these changes in atmospheric processes affect the downward solar radiation at the surface, as directly calculated by the climate models using their physical parameterization schemes.

For solar energy applications, it is relevant to know not only the changes in the seasonal mean solar radiation but also how these changes are distributed through different times of the day. The analysis of the latter requires subdaily data of solar radiation, which were archived only by a small number of modeling groups in CMIP. Nevertheless, with the limited data, we will make a first attempt to quantify the trends at different times of the day.

8.2 DATA FROM CLIMATE MODEL SIMULATIONS

We will analyze two sets of climate model simulations from the CMIP3 archive (http://www-pcmdi.llnl.gov/ipcc/about_ipcc.php). Although a

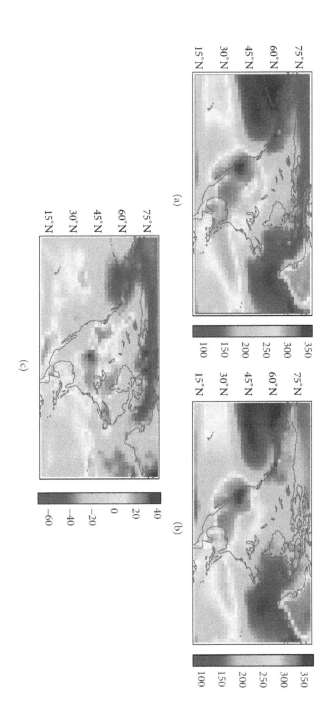

Figure 1: (a) The July climatology of downward shortwave radiation at the surface over North America from the 21st century run with GFDL CM2.0, based on 2080–2100 average. (b) The same as (a) but from the 20th century run, based on 1980–2000 average. (c) The trend, defined as (a) minus (b). The color scales with units in W/m2 are shown on the right.

large collection of model outputs have been archived by CMIP3 and the more recent CMIP5 (Taylor et al. [7]), the majority of the data are monthly means. Subdaily outputs are archived by only a small number of modeling groups and for short time periods. In this study, we choose to analyze the GFDL CM2.0 and MRI CGCM2.3.2 simulations in CMIP3. Both groups provided subdaily (3-hourly) archives of downward shortwave radiation at the surface for selected time slices in late 20th century and late 21st century. The surface radiation budget for CMIP3 20th century simulations was analyzed and compared to observation by Wild [8]. Over the midlatitude belt from 30 N to 60 N, the MRI model has about +15 W/m² bias while the GFDL model has about −5 W/m² bias in the all-sky downward shortwave radiation at the surface (see Figure 6 in Wild [8]).

For both models, the GHG-induced trend will be deduced from the difference between the SRES A1B run (with increasing GHG concentration according to the A1B scenario) for the 21st century and the 20C3M run for the 20th century. We will first use the monthly mean archives to calculate the centennial trend, defined as the climatology of 2080–2100 minus the climatology of 1980–2000. For the analysis of the trends at different times of the day, we will use the 3-hourly model outputs as available from CMIP3. Each modeling group only provided the high-frequency data for short time slices in the late 20th century and late 21st century. Based on the availability of data, we will use the difference between year 2100 (from SRES A1B runs) and year 2000 (from 20C3M runs) to calculate the trends at selected local times during the day in North America.

This study will focus on the climate trend over North America. Over the Eurasian continent, Meleshko et al. [9] have shown that CMIP3 models project a 2–4% increase in winter and 2–10% decrease in summer of cloud cover (total cloud fraction) over Russia. The latter corresponds to an increase of 4-5 W/m² in summer in the area-averaged downward shortwave radiation at the surface. Since the populous regions in North America are located at lower latitudes than Russia, a similar change in cloud cover over North America is expected to produce a greater change in the downward shortwave radiation at the surface.

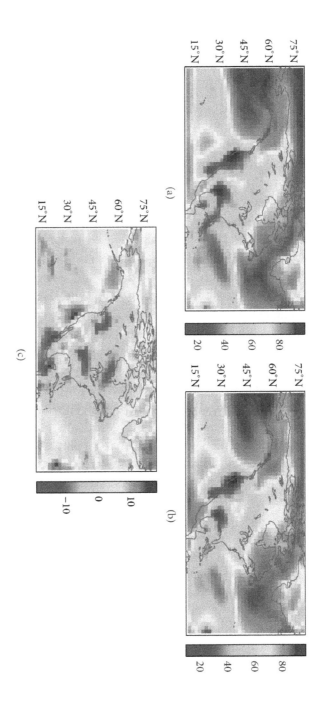

FIGURE 2: (a) The July climatology of total cloud fraction from the 21st century run with GFDL CM2.0, based on 2080–2100 average. (b) The same as (a) but from the 20th century run, based on 1980–2000 average. (c) The trend, defined as (a) minus (b). The color scales with units in percentage are shown on the right.

8.3 RESULTS AND DISCUSSION

8.3.1 MONTHLY MEAN CLIMATOLOGY AND TREND

Figures 1(a) and 1(b) show the July climatology of the downward short-wave radiation at the surface over North America from GFDL CM2.0 simulations. Figure 1(a) is the 2080–2100 average from the SRES A1B run and Figure 1(b) is the 1980–2000 average from the 20C3M run. Figure 1(c) shows the trend deduced from the difference (future minus present) between Figures 1(a) and 1(b). For the climatology in Figures 1(a) and 1(b), although the downward solar radiation is zonally uniform at the top of the atmosphere, it becomes significantly nonuniform upon reaching the surface. This longitudinal nonuniformity is strongly influenced by cloudiness, as can be readily seen in Figure 2, the counterpart of Figure 1 for the total cloud fraction in July from the same simulations. For example, the strong downward shortwave radiation at the surface over Western and Southwest USA corresponds to the mostly clear-sky condition in July in those regions. In Figure 1(c), the model projected an overall positive trend over most of the United States, except for a small area in the Southwest USA. The increased sunlight at the surface, on the order of about 20 W/m^2 (over the 21st century) or close to 10% of the climatology, is related to a reduced cloudiness over the USA (Figure 2(c)).

Figure 3 is similar to Figure 1 but for January. The trend in winter (Figure 3(c)) is an increase of the downward shortwave radiation over Southern and Southwest USA and a decrease of it over the northern half of the USA. The increased sunlight over Southwest USA in the cold season is consistent with the known projection by climate models of a drying trend in that region (Seager et al. [5] and Baker and Huang [6]). This drying trend is related, in part, to the poleward shift of storm tracks (Seager et al. [5]), which is consistent with the decrease in sunlight over Northern USA since the increase in storm activities implies an increase in cloudiness.

The downward shortwave radiation at the surface simulated by MRI CGCM2.3.2 is shown in Figure 4. For conciseness, only the trends are shown. Figures 4(a) and 4(b), for July and January, respectively, are the counterparts of trends in Figures 1(c) and 3(c). In summer, the trend simulated by the MRI model is significantly different from that produced by the GFDL model. While both models project an increase of sunlight over Eastern USA and the Pacific Northwest, MRI projects a neutral to slightly negative trend over Western USA compared to a positive trend by the GFDL model. The trend in winter is more robust. Both models produced a decrease of sunlight over Northern USA and an increase of it over Southern USA. For both models, the magnitude of these trends is on the order of 10 W/m², or about 10% of the January climatology. As a notable difference, over the Southwest USA, GFDL CM2.0 produced a positive trend while MRI CGCM2.3.2 produced a neutral to negative trend. Since the trend in the shortwave radiation is strongly influenced by the trend in cloudiness which is highly parameterized in climate models, the differing projections by the two models are not surprising. A better agreement in the trend is found in January, possibly because the decrease in sunlight over Northern USA in the cold season is related to the poleward shift of storm tracks under an increasing GHG forcing, a phenomenon that is large scale in nature and is robustly simulated by the majority of climate models in CMIP3 (e.g., Yin [10]).

8.3.2 TRENDS AT DIFFERENT TIMES OF THE DAY

We next use the more limited data of 3-hourly model outputs to deduce the trends in the downward shortwave radiation at the surface as a function of the times of the day. Both GFDL CM2.0 and MRI CGCM2.3.2 provide the 3-hourly archives for the downward shortwave radiation for the year 2000 (from the 20C3M runs) and 2100 (from the SRES A1B runs). Figure 5 illustrates a 3-hourly time series of the downward shortwave radiation at the surface averaged over multiple grid points in Northern Arizona, for the year 2000, from the 20C3M simulation with GFDL CM2.0. The trend (now defined as the 2100 average minus the 2000 average) discussed in this section will be for a specific time of the day, averaged over either July or January.

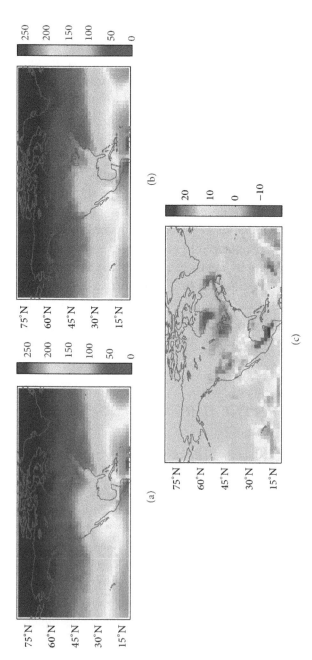

FIGURE 3: (a) The January climatology of downward shortwave radiation at the surface from the 21st century run with GFDL CM2.0, based on 2080–2100 average. (b) The same as (a) but from the 20th century run, based on 1980–2000 average. (c) The trend, defined as (a) minus (b). The color scales with units in W/m² are shown on the right.

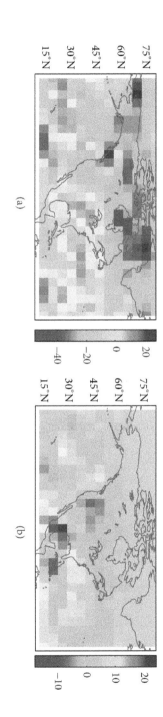

(a)

(b)

FIGURE 4: (a) The trend of the downward shortwave radiation at the surface for July from the MRI CGCM2.3.2a simulations. (b) The same as (a) but for January. The trend is defined as the 2080–2100 mean climatology minus the 1980–2000 climatology. Color scales with units of W/m² are shown on the right.

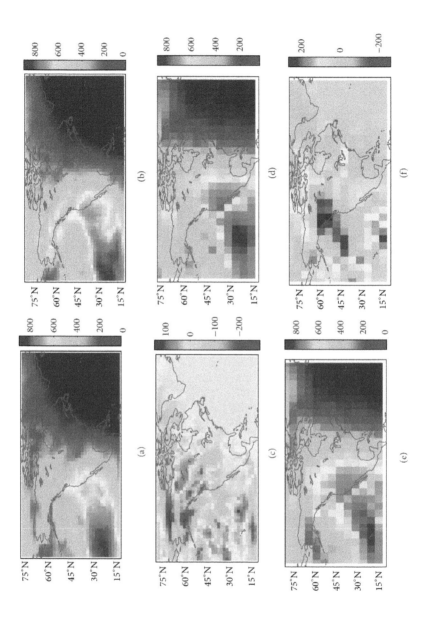

FIGURE 5: The time series of three-hourly downward shortwave radiation at surface, averaged over multiple grid points that cover Northern Arizona, for the year 2000 from the GFDL CM2.0 20th century simulation. The unit on the ordinate is W/m^2.

Figures 6(a)–6(c) are similar to Figures 1(a)–1(c) but for the GFDL CM2.0 simulations of the shortwave radiation at 3 PM local time of US West Coast. Figure 6(a) is the average over July 2100 and Figure 6(b) the average over July 2000. The trend, defined by 2100 minus 2000, is shown in Figure 6(c). Figures 6(d)–6(f) are the counterparts of Figures 6(a)–6(c) but for the MRI CGCM2.3.2 simulations. Figure 7 is similar to Figure 6 but for the shortwave radiation at 9 a.m. local time of US West Coast. Note that the climatological values and trends in Figures 6 and 7 are much higher than the monthly mean values in Figure 1 because the latter are the average over the whole day, including nighttime. Just like the monthly mean, in July, the trends projected by the two models are significantly different. Also, within each model, the trend at a particular time of the day is different from the trend of the monthly mean which includes the contributions from all times of the day.

While the trends in July are not robust, the trends in January as shown in Figure 8 exhibit a greater degree of consistency between the two models and across different times of the day. For conciseness, only the trends are shown. Figures 8(a) and 8(b) are the January trend at 3 p.m., and Figures 8(c) and 8(d) are the January trend at 9 a.m. local time of US West Coast. The top row (Figures 8(a) and 8(c)) is from GFDL CM2.0 and bottom row (Figures 8(b) and 8(d)) from MRI CGCM2.3.2 simulations. The decrease in sunlight at the surface over Northern USA and the increase of it over Southern USA, as previously shown in the monthly mean plots in Figures 3 and 4, can be identified in all four panels in Figure 8. From Figures 8(c) and 8(d), at the time when the entire United States is in the middle of the day, the January trend of the decrease in solar radiation in Northern USA can locally reach $100\,W/m^2$, or slightly over 10% of the local climatological value at that time.

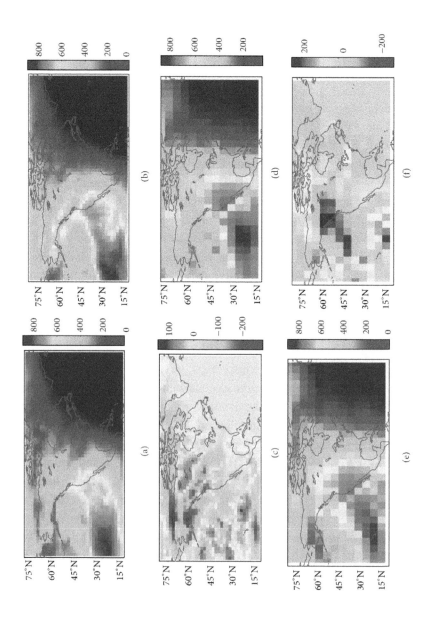

FIGURE 6: (a) The downward shortwave radiation at the surface at 3 PM local time of US West Coast, averaged over July 2100, from the GFDL CM2.0 21st century run. (b) The same as (a) but for the average over July 2000 from the GFDL CM2.0 20th century run. (c) The trend, defined as (a) minus (b). Panels (d)–(f) are the counterparts of (a)–(c) from the MRI CGCM2.3.2 simulations. The color scale with units of W/m^2 is shown on the right for each panel.

8.4 CONCLUDING REMARKS

The most robust finding of this study is the wintertime (January) trend in the downward shortwave radiation at the surface over the United States. It exhibits a simple pattern of a decrease of sunlight over Northern USA and an increase of sunlight over Southern USA. This structure is simulated by both GFDL and MRI models and can be identified even at different times of the day. It is broadly consistent with the known poleward shift of storm tracks in the cold season in climate model simulations under an increasing GHG forcing. The negative trend in Northern USA is more prominent. Quantitatively, the centennial trend of the downward shortwave radiation at the surface in that region is on the order of 10% of the climatological value for the monthly mean (averaged over all times of the day) and slightly over 10% at the time when it is midday in the United States. This indicates a nonnegligible influence of the GHG forcing on solar energy in the long term. Nevertheless, when dividing the 10% by a century, in the near term, the impact of the GHG forcing is relatively minor such that the estimate of solar power potential using present-day climatology will remain useful in the coming decades. The global climate models used in this study have relatively coarse resolutions with the horizontal grid size exceeding 100 km. For the assessment of solar power potential at a specific site of existing or future solar power plant, it will be desirable to perform climate downscaling (e.g., Mearns et al. [11] and Sharma and Huang [12]) to take into account the effects of small-scale topography on cloudiness. The findings of this work will serve as a useful reference for future studies in that direction.

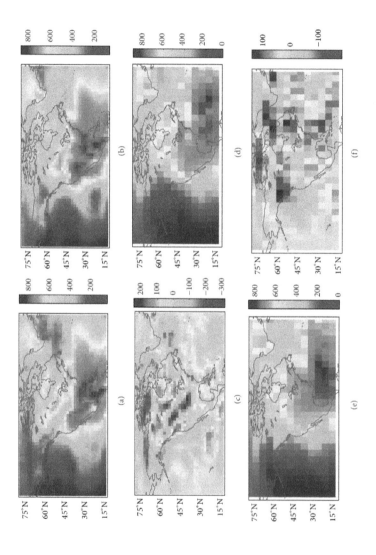

FIGURE 7: The same with Figure 6 but for the downward shortwave radiation at 9AM local time of US West Coast for July. Panels (a)–(c) are from GFDL CM2.0 and panels (d)–(f) are from MRI CGCM2.3.2 simulations. The color scale with units of W/m² is shown on the right for each panel.

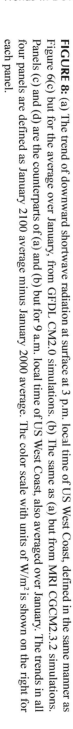

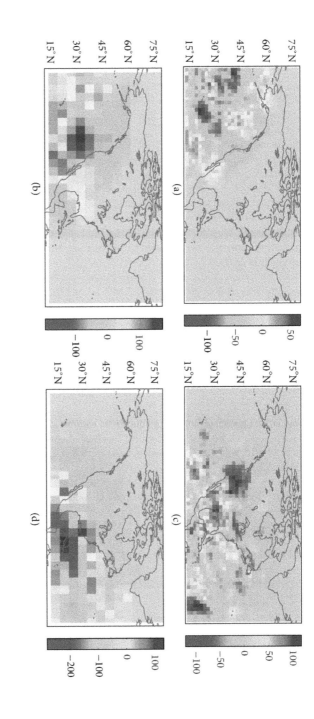

FIGURE 8: (a) The trend of downward shortwave radiation at surface at 3 p.m. local time of US West Coast, defined in the same manner as Figure 6(c) but for the average over January, from GFDL CM2.0 simulations. (b) The same as (a) but from MRI CGCM2.3.2 simulations. Panels (c) and (d) are the counterparts of (a) and (b) but for 9 a.m. local time of US West Coast, also averaged over January. The trends in all four panels are defined as January 2100 average minus January 2000 average. The color scale with units of W/m² is shown on the right for each panel.

REFERENCES

1. Z. Li, H. W. Barker, and L. Moreau, "The variable effect of clouds on atmospheric absorption of solar radiation," Nature, vol. 376, no. 6540, pp. 486–490, 1995.
2. E. Maxwell, R. George, and S. Wilcox, "A climatological solar radiation model," in Proceedings of the Annual Conference, American Solar Energy Society, p. 6, Albuquerque, NM, USA, 1998.
3. R. George and E. Maxwell, "High-resolution maps of solar collector performance using a climatological solar radiation model," in Proceedings of the 1999 Annual Conference, p. 6, American Solar Energy Society, Portland, Me, USA, 1999.
4. G. A. Meehl, C. Covey, K. E. Taylor, et al., "The WCRP CMIP3 multimodel dataset: A new era in climatic change research," Bulletin of the American Meteorological Society, vol. 88, no. 9, pp. 1383–1394, 2007.
5. R. Seager, M. Ting, I. Held et al., "Model projections of an imminent transition to a more arid climate in southwestern North America," Science, vol. 316, no. 5828, pp. 1181–1184, 2007.
6. N. C. Baker and H.-P. Huang, "A comparative study of precipitation and evaporation in semi-arid regions between the CMIP3 and CMIP5 climate model ensembles," Journal of Climate, vol. 27, pp. 3731–3749, 2014.
7. K. E. Taylor, R. J. Stouffer, and G. A. Meehl, "An overview of CMIP5 and the experiment design," Bulletin of the American Meteorological Society, vol. 93, no. 4, pp. 485–498, 2012.
8. M. Wild, "Short-wave and long-wave surface radiation budgets in GCMs: a review based on the IPCC-AR4/CMIP3 models," Tellus A, vol. 60, no. 5, pp. 932–945, 2008.
9. V. P. Meleshko, V. M. Kattsov, V. A. Govorkova, P. V. Sporyshev, I. M. Shkol'nik, and B. E. Shneerov, "Climate of Russia in the 21st century. Part 3. Future climate changes calculated with an ensemble of coupled atmosphere-ocean general circulation CMIP3 models," Russian Meteorology and Hydrology, vol. 33, no. 9, pp. 541–552, 2008.
10. J. H. Yin, "A consistent poleward shift of the storm tracks in simulations of 21st century climate," Geophysical Research Letters, vol. 32, no. 18, Article ID L18701, pp. 1–4, 2005.
11. L. O. Mearns, R. Arritt, S. Biner et al., "The north american regional climate change assessment program: overview of phase I results," Bulletin of the American Meteorological Society, vol. 93, no. 9, pp. 1337–1362, 2012.
12. A. Sharma and H.-P. Huang, "Regional climate simulation for Arizona: impact of resolution on precipitation," Advances in Meteorology, vol. 2012, Article ID 505726, 13 pages, 2012.

PART IV

HYDROPOWER AND THE WEATHER

CHAPTER 9

The Impact of Droughts and Climate Change on Electricity Generation in Ghana

EMMANUEL OBENG BEKOE AND FREDRICK YAW LOGAH

9.1 INTRODUCTION

Hydrological drought which is considered in this paper is when the water reserves available in the dam/reservoir fall below the statistical average. Hydrological drought tends to show up more slowly because it involves stored water that is used but not replenished. The Volta River Authority (VRA) was established under the Volta River Development Act 1961 (Act 46) with the objective to develop the hydroelectric potential of the Volta River for supply of electric energy for industrial, commercial and domestic use in Ghana and to some neighboring countries. The principal functions of VRA were to generate electric power,

The Impact of Droughts and Climate Change on Electricity Generation in Ghana. Bekoe EO and Logah FY. Environmental Sciences *1,13–14 (2013). The work is made available under the Creative Commons Attribution License, http://creativecommons.org/licenses/by/3.0/.*

initially, by the construction of a dam and hydroelectric generating station at Akosombo, and to construct and operate a transmission system to carry the power to serve industrial, commercial and domestic needs of the country. VRA's generation activities cover the operation of two hydroelectric plants one in Akosombo (912 MW) and the other at Kpong (160 MW), also on Volta River, downstream from Akosombo. In addition VRA runs a 30 MW diesel generating station at Tema, which was commissioned in 1992. In 1995, the VRA started constructing a new 330 MW Combined Cycle Thermal generating Plant, comprising two 110 MW Combustion Turbines and one 110MW Steam Turbine Generator and associated Heat Recovery Steam Generator (HRSG) at Abaodze, near Takoradi. The Thermal Project at Aboadze is bringing on board about 660 MW of power. The Asogli Thermal Unit (a private power producer) power plant brings about 12% power. In all electricity from hydro is about 60% against thermal of 40% (PURC, 2011). VRA has been exchanging electrical power with its Ivorian counterparts Energie Electrique de la Cote d'Ivoire (EECI) and Compagnie Ivorienne d'Electricite (CIE) since February 1984. Currently however, VRA is a net importer of electricity from CIE.

Since VRA's establishment, water shortage due to droughts or low rainfalls have befallen the Authority in some years. In the year 2007, Ghana experienced the severest electricity power rationing, equating to about 24hrs light in 48 hrs. This rationing was attributed at the time to the low water level in the Volta Lake which harbours the Akosombo and the Kpong Dams due to poor rainfalls in the Volta Basin. Nearly all industries in the Country suffered production problems culminating in some folding up. This paper presents some of the effects of drought which is an extreme weather phenomenon on the operations of hydroelectricity generation in Ghana. It does so by analyzing rainfalls in the basin and lake (dam water levels at the intake point) water levels of the Akosombo dam in the Volta Basin to establish whether in reality, the main causes of the water shortage (hydrological drought) were due to drought and whether as a country we have the expertise to forecast drought.

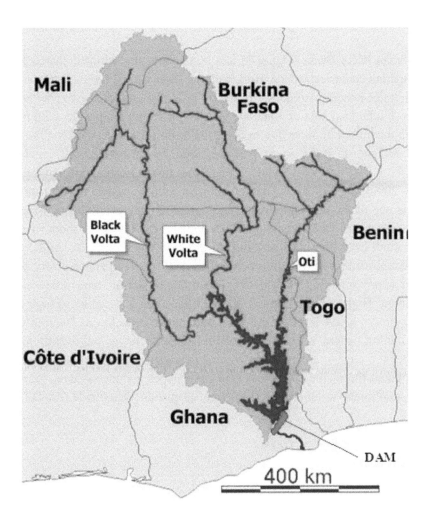

FIGURE 1: The Volta River basin (shared by Ghana Ivory Coast, Burkina Faso, Togo, Benin and Mali)

9.2 MATERIALS AND METHODS

9.2.1 LOCATION

The Volta River Basin (Figure 1) is a trans-national catchment shared by six riparian countries: Ghana, Burkina Faso, Togo, Cote d'Ivoire, Benin and Mali. It lies between latitudes 5°N and 14°N and longitudes 2° E to 5°W and drains a total land area of 400,000 km² . About 15 million people, with per capita income of $650/yr live in the Volta basin. The watershed is 42% in Burkina Faso, 40.2% in Ghana, 6.35% in Togo, 4.57% in Mali, 3.62% in Benin, and 3.24% in Cote d'Ivoire (Green Cross International, 2001; Andreini et al., 2000). The Volta Lake has a maximum submerged area of 8500 km² , and is one of the most important physiographic features in Ghana. Major tributaries to the lake include the Black Volta, the White Volta (both rivers flow south from Burkina Faso), the Oti and the Daka rivers. The landscape is predominantly flat with elevations below 1000 m. Temperatures vary between approximately 16°C and 40°C depending on season, time of day, and elevation. Highest rainfall levels occur in the south and can be as high as 2000 mm/yr, where levels in the driest regions can be as low as 200 mm/yr. The rainfall pattern is usually unreliable, while declining soil fertility and accelerated erosion are evidence of land degradation within the basin and throughout the region. The main land cover types include savanna, grassland, rain forest, water bodies, shrubs and croplands (Oguntunde, 2004).

9.2.2 DATA COLLECTION

Rainfall figures for the period 1971 to 2007 were collected from six stations (Figure 2) namely Bole, Kete-Krachi, Tamale, Yendi, Wa and Navrongo representing the Volta Basin at Akosombo together with dam water levels spanning 1970-2005 were assessed for this study from the Ghana Meteorological Agency and VRA respectively. Stations in Burkina Faso and Mali were not considered because availability was a problem and more so, with the construction of an irrigation dam in 1995 at Bagri in Burkina Faso the flows from this area is controlled except during spilling which normally occurs in the wet season.

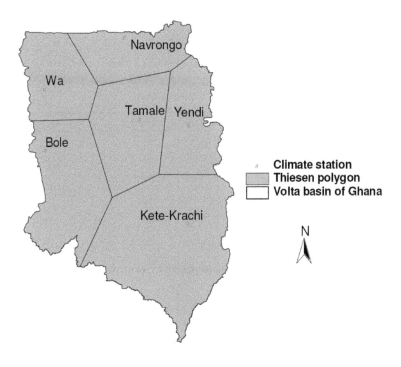

FIGURE 2: Map showing the rainfall stations used for the computation.

9.2.3 DATA ANALYSIS

With the close relationship between water availability and drought, analysis of rainfall, drought and water levels was carried out to establish whether occasional water shortage experienced in Ghana in the years 1983, 1993/94, 1998 and, 2006 and the attendant rationing of hydro electricity power (VRA, 2011) for those years were really droughts years. The Thiessen Polygon method was used to weight the six rainfall stations to compute the annual areal rainfall (Table 1).

TABLE 1: Rainfall station weights in the Volta Basin

Station	Fraction contributed to rainfall Percentage	Total Contribution
Kete-Krachi	0.3	32.4
Tamale	0.2	15.6
Yendi	0.1	11.7
Bole	0.2	18.2
Navrongo	0.1	10.6
Wa	0.1	11.5
	1.0	100.0

The annual area rainfall distribution graph is shown in Figure 3 and that of the water levels in the dam at Akosmbo shown in Figures 4&5. Due to the non-continuous nature of rainfall in time and space, its statistical description can be quite complex. A drought threshold is an essential element in categorizing the drought events in drought analysis. Drought threshold is a constant demand where the droughts are defined as periods during which the discharge is below the threshold level (Fadhilah Yusof and Foo Hui-Mean, 2012; Fleig et al., 2006). A threshold that applied commonly in monitoring rainfall and preparing drought alerts is at the seventieth percentile of the rainfall when the rainfall in a certain period is less than seventy percent of normal precipitation (FAO, 2003). Following, the Probability of Exceedence (POE) method was used for analysis. In determining the threshold values, the rainfall figures were sorted in ascending order where the value at the seventy percent of the rainfall data is recognized as the seventieth percentile of the rainfall series, p70 (POE of 30). The drought rainfall threshold (FAO, 2003) is then set according to equation (1)

$$PN/100 = L \tag{1}$$

P is the selected percentile and N is the total number of measurements in the data set. The threshold is the Lth value of amount of rainfall.

The recurrence or Return period of the drought T is also computed using the method

$$\text{Return Period (T)} = (n+1)/m \tag{2}$$

Where n is the number of years considered and m is the rank of the rank of the event being considered

9.3 RESULTS

For this analysis conservative values of threshold between the Probability of Exceedence (POE) 33% to 20% was considered a dry year but no drought, POE between 20% and 10% is dry and nearing drought with POE (10%) and below considered a very drought year because Ghana seldomly experiences drought. These POE's were computed and is also shown in figure 3.

From figures 3 it could be deduced that the years 1983, 1992 and 2001 were actual drought years in Ghana. On the other hand 1977, 1988, 1993 and 2006 fell within the POE (20%) indicating that they were nearing drought years.

From the water levels plot (Figures 4 & 5) it is established that the dam fills up from July–October, and commences drawdown from November

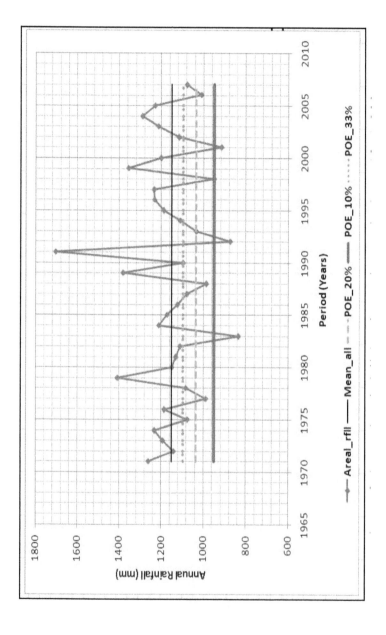

FIGURE 3: Annual Rainfall for the Volta Basin from 1971-2007

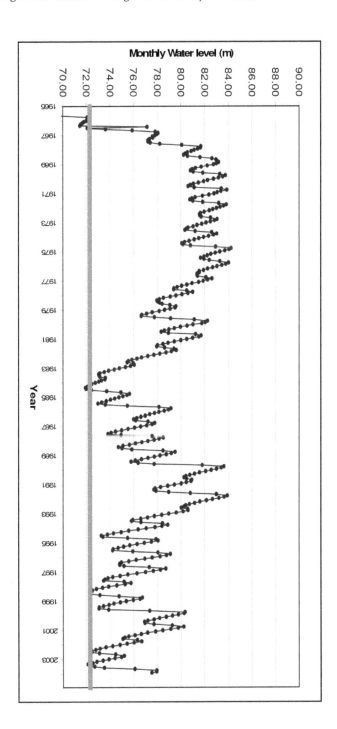

FIGURE 4: Graph showing monthly water levels of Akosombo Dam (1966-2003) {Courtesy, VRA}

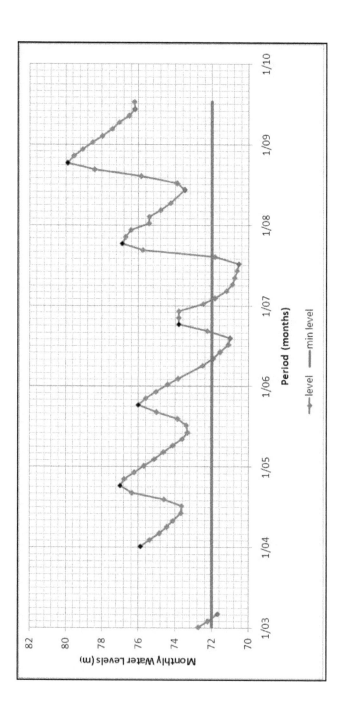

FIGURE 5: Graph showing monthly water levels of Akosombo Dam (Jan 2003-Jul, 2009)

–June. According to VRA, the designed minimum drawdown of the impounded water for electricity generation is 72 meters. Yet in Figures 4&5, the years 1983, 1998, 2003, 2006 and 2007 had the recorded water levels falling below the minimum level for power generation indicating that the inflows into the dam was not adequate.

In Table 2, comparing rainfall in the nearing drought and drought years and electricity rationing years, it can be said that the power rationing years of 1983/84, 1998, 2006/07 were actually due to drought. However, the rationing year of 2003 cannot be wholly explained as a drought year. This may be due to effects of previous drought years of 2001 and below average rainfalls in 2002.

TABLE 2: Comparison of drought years and rationing years

Rainfalls nearing drought and drought years	Electricity rationing years
1997, 1983, 1988, 1992, 1998, 2001, 2006	1983/84, 1998, 2003, 2006/07

The computed hydrological drought using equation (2) shows a return period of 10 years from the 37 years of rainfall indicating that hydrological drought may return every 10 years.

9.4 DISCUSSION

9.4.1 SOCIOECONOMIC CONSEQUENCE OF LOW INFLOWS

The 2006/2007 electricity power rationing reduced power supply to households and Industry to about 24hrs light in 48 hrs and was the severest ever witnessed in Ghana with catastrophic consequences. Out of six turbines, only two were in operation churning about 400MW out of 1180MW from the hydropower generation unit. This affected all sectors of the economy including industry, mining and domestic. Manufactures reduced output, shortened the workweek and invested heavily in power generators. Drought left factories idle and broad swaths of Accra and other cities dark at night with the most feared result that years of efforts spent creating an image of Ghana as receptive to foreign investment could be jeopardized. Databank, a brokerage firm in Ghana estimated at the time that the outages were forcing companies to spend \$62 million a month, or about \$744 million a year, on extra power generation, or about 6% of the country's entire economic output. DatabankGhana forecasted at the time that the power shortage cut 2007 economic growth from 6.5% to between 4% and 5% (Philips, 2007). The Statesman Newspaper of 19/2/2007 featuring an article titled "Power rationing cost Ghana ¢140bn in taxes" extolled. Further, the commissioner of Internal Revenue Service at the time disclosed at Akosombo that the power rationing programme made the Internal Revenue Service to lose revenue estimated at ¢140 billion (US\$ 14 Million) which could have been collected as taxes for Government in 2006. This, the Commissioner said was so because the power cut exercise made many companies in the country to record low earnings which affected taxes. In another comment on energy rationing in 2006 the Chief Executive of the Ghana Chamber of Mines reported on VOA news (VOA, 2006) that "Cutting back by 50 percent means that, it is almost like cutting back production by 50 percent, because, although we have installed capacity for selfgeneration, it's extremely expensive...," "It will mean getting regular supply of diesel, at the cost that we get it will perhaps mean generating power at 15 cent per kilowatt hour, which is almost three times what we get from VRA" (VOA, 2006). Also commenting on the sufferings of Ghanaians on the power rationing, Dr. Kwabena Anaman, director of research at Ghana's Institute of Economic Affairs said "if the situation persists, growth rates set out in this year's government budget will not be achieved. He says reliable energy is vital to the development of the country's economy" (VOA, 2006).

9.4.2 DROUGHTS AND CLIMATE CHANGE EFFECTS ON THE HYDROPOWER GENERATION

Climate has a great impact on ecosystems and hence the livelihoods of populations that depend on these systems (Boko et. al., 2007). Africa is one of the most vulnerable continents to climate change and variability.

Lars (2006) attributed this high vulnerability of Africa to the following four reasons:

- Much of Africa, particularly Sub Sahara African, is in dry or sub humid agroecological zones and recent climate change models (GCMs) show that climate change will affect the rainfall patterns in these zones.
- A very large proportion of the African population, especially those in SubSahara Africa, depends on natural resources for food and income. Therefore, changes in climate that impact these resources will have serious effects on the livelihoods of a vast majority of the African population.
- High levels of poverty and an already very high pressure on Africa's natural resource base means people have very little capacity to adapt to climate change.
- The continuous degradation of natural resources in the continent, such as the progressive degradation of the agricultural lands, forests and savannahs due to unsustainable resource management practices, results in decreased natural capacity of these resources to adjust to changes in climate.

Ghana is well endowed with water resources, but the amount of water available changes markedly from season to season as well as from year to year (WRI, 2000). Droughts; one of the consequences of decreasing rainfalls as shown in this paper could be catastrophic to the nation. In August and September of 2007, the same year that the serious power cuts occurred, the basin saw widespread and devastating floods due to extreme rainfall from July-September. These floods displaced hundreds of thousands of people, particularly in the three northern regions, with resettlement and other mitigating costs estimated to be in the millions of dollars. This was an indication that the climate may be experiencing changes.

Literature such as WRI, (2000), Andah et al. (2004) Kuntsmann and Jung, (2005) and WRI, (2010) shows there is evidence of climate change effects on Ghana and its water resources. Noting that climate change in Ghana is a certainty, in whatever form the change may come hydropower generation could be affected one way or the other. If there is more precipitation than normal the dam could be put at risk because of structural problems and if it is less rainfall, then drought as evidenced in this paper could result.

9.5 CONCLUSION

Considering the thirty seven years of data used for this paper and four drought years identified, the return period for drought in this basin is 10 years. This paper establishes that 1983, 1998, 2006 were actually drought years and therefore the power rationing for the ensuing years of 1983-84, 1999, 2006-7 were deserving. However, the rationing year of 2003 could not be explained as a drought year. This may be due to effects of previous drought years of 2001 and below average rainfalls in 2002.

If socioeconomic consequences as a result of drought are to be mitigated there is the need to manage the hydropower dam and the hydrology of the basin sustainably to forestall many of the consequences such as loss of revenue, power rationing which may affect industry adversely and attendant socio economic problems.

It may be true that the traditional mainstays of Ghana's economy depend relatively little on electricity in the pre 2000 years. However, with the widening of the electricity coverage to most part of the country in the late 1990's and after the year 2000, this assertion may be faulted. From the above analysis vis a vis the socio economic problems encountered, Ghanaians acknowledge that there is the need for power sufficiency as many agree that industry must lead the way to the future if Ghana is to escape poverty. To industrialize means to build new industries which requires vital foreign investment. Though the Ghanaian business people are used to difficult situations like power rationing, investors looking to invest in Ghana, seeing problems like unreliable power supply due to low rainfalls and drought may reconsider their investment.

Thus there is the need to improve forecasting of extreme weather phenomenon such as droughts so that the consequences of this water shortage could be foreseen and curtailed. Adaptation measures could be employed if the problem is climate change. Fortunately the Government of Ghana has brought on board independent power generators who are contributing nearly 40-50 of present power supply in Ghana.

REFERENCES

1. FAO (Food and Agriculture Organization), A Perspective on Water Control in Southern Africa, Land and Water Discussion Paper 1, Rome, 2003, ISSN 1729-0554.
2. F. Yusof and F. Hui-Mean, Use of Statistical Distribution for Drought Analysis. Applied Mathematical Sciences, Vol. 6, 2012, no. 21, 1031 – 51
3. Green Cross International, Burkina Faso, Trans-boundary Basin Sub-Projects: The Volta River Basin. 2001. Website: www.gci.ch/GreenCrossPrograms/waterres/pdf/WFP_Volta Accessed on 2010-12-14.
4. H. Kuntsmann. and G. Jung, Impact of regional climate change on water availability in the Volta basin of West Africa. In: Regional Hydrological Impacts of Climatic Variability and Change (Proceedings of symposium S6 held during the Seventh IAHS Scientific Assembly at Foz do Iguacu, Brazil, April (2005). IAHS Publ. 295 http://www.waterconserve.info/shared/reader/welcome.aspx?linkid=105432&keybold=deforestation%20flooding. Accessed on 2011-10-30.
5. L. Hein. Climate Change in Africa. (2006) URL: http://www.cicero.uio.no/fulltext/index_e.aspx?id=5249. Accessed on 2008-05-21.
6. M., I. Boko, A. Niang, C. Nyong, A. Volgel, M. Githeko, B. Medany, R. Osman-Elasha, R. Tabo and P. Yanda,: Africa. Climate change 2007: Impacts, Adaptation and Vulnerability. Contribution of Working Group II to the Fourth Assessment Report of the Intergovernmental Panel on Climate Change, M.L. Parry, O.F. Canziani, J.P. Palutikof, P.J. van der Linden and C.E. Hanson, Eds., Cambridge University Press, Cambridge UK, 433-467. 2007
7. M. Andreini, N. van de Giesen, A. van Edig, M. Fosu, W. Andah, Volta Basin water balance. ZEF-Discussion Papers on Development Policy, Number 21, Center for Development Research, ZEF, Bonn, Germany, 2000.
8. M. M. Philips, How Ghana's Economic Turnaround Is Threatened. An article in the Wall Street Journal of August 6th 2007. (http://online.wsj.com/article/SB118635544536388723.html#) Assessed on 15h September 2009.
9. P.G. Oguntunde, Evapotranspiration and complementarity relations in the water balance of the Volta Basin: field measurements and GIS-based regional estimatesEcology and Development Series, vol. 22.Cuvillier Verlag, Gottingen, p. 169, 2004
10. PURC, Public Utility Regulatory Commission of Ghana, 2011.
11. WRI-CSIR, Climate Change Effects on Hydrology and Water Resources and Adaptation strategies in Ghana. UNESCO Funded Project, WRI Technical Report. Ghana, 2010.
12. VOA (2006) Energy Crisis Impacts Ghana's Mining Industry (VOA news). http://www.voanews.com/english/archive/2006-09/2006-09-12-voa36.cfm?moddate=2006-09-12) Assessed 15th September, 2009.
13. VRA (2011). Home page for VRA www.vra.com assessed September, 30, 2011

14. W.E.I. Andah, N. van de Giesen and C. A. Biney, Water, Climate, Food, and Environment in the Volta Basin. Contribution to the project ADAPT (Adaptation strategies to changing environments), 2004.

15. WRI-CSIR, Climate change vulnerability and adaptation assessment on water resources of Ghana. A UNFCC/EPA/WRI Accra report, 2000.

CHAPTER 10

Tailoring Seasonal Climate Forecasts for Hydropower Operations

P. BLOCK

10.1 INTRODUCTION

Seasonal climate forecasting capabilities continue to advance, attributable predominantly to enhanced observations, computing power, better physical understanding of the climate system, and experience (Barnston et al., 1994, 2005; Goddard et al., 2003). Their principle goal is to reduce climate-related risks, providing advance information to potentially improve decision-making and increase societal benefits, especially over the long term. Currently, however, there exists little evidence of direct forecast use in operations, especially in water resources management, even in regions of scarcity. This is often ascribed to water managers tendency to act in a risk averse manner, "poor" forecast skill or scale, difficulty in integrating

Tailoring Seasonal Climate Forecasts For Hydropower Operations. © *Block P.* Hydrology and Earth System Sciences *15 (2011). doi:10.5194/hess-15-1355-2011. Licensed under Creative Commons Attribution 3.0 Unported License, http://creativecommons.org/licenses/by/3.0/.*

forecasts into existing decision support systems, lack of focus on specific user needs, anticipated shifts in the water sector, management and political disincentives, individual and institutional inflexibility, behavioral effects, and informational constraints (Pulwarty and Redmond, 1997; Hamlet et al., 2002; Ritchie et al., 2004; Rayner et al., 2005; Broad et al., 2007; Johnston et al., 2007; Lemos, 2008; Millner, 2009; Ziervogel et al., 2010).

The abundance of research and literature over the past decade identifying challenges and impediments should act as a stimulus for case studies evaluating potential economic benefits and improved reliability through forecast inclusion. These two determinants are powerful motivators for water resources managers and policy makers, and forecastinduced positive outcomes may provide incentive to address other barriers. Previous research studies have advocated for demonstrations of such effective forecast use (e.g. Pagano et al., 2001). Minimal applications within the water resources community, however, seek to quantify the actual monetary and reliability gains or losses of including a forecast in comparison to commonly accepted climatology-based operations (i.e. based on average climate conditions), and most of those examples refer only to perfect forecasts, excepting a few (e.g. Yeh et al., 1982; Kim and Palmer, 1997; Yao and Georgakakos, 2001; Hamlet et al., 2002; Chiew et al., 2003; Maurer and Lettenmaier, 2004; Axel and Ceron, 2007; Sankarasubramanian et al., 2009). An absence in forecast adoption is unmistakable (Rayner et al., 2005), and further exaggerated in developing countries with limited hydrologic observations (Patt et al., 2007; Ziervogel et al., 2010).

This motivates the current research to demonstrate the improved economic value and reliability resulting from a flexible seasonal climate forecast—hydropower modeling system, given biophysical, policy, and economic constraints, by mitigating losses and capitalizing on opportune conditions (Hellmuth et al., 2007). Forecasts, coupled with flexible operating rules, may lead to optimal or more efficient reservoir management of storage and release volumes based on expected probabilistic future conditions (Karamouz and Vasiliadis, 1992; Faber and Stedinger, 2001). Gaining an understanding of expectations from a realistic, imperfect forecast imbedded in a dynamic operational system could prove enticing for water managers to adopt forecast inclusion, or justification for rejecting. Appreciating benefits and reliability in a context of climate variability begins to address a number of the aforementioned impediments (summarized well in Ziervogel et al., 2010).

This paper commences with a description of the application site, the Blue Nile basin in Ethiopia's highlands, followed by an outline of the linked forecast-hydropower modeling system. The economic value and reliability produced from the seasonal climate forecast driven system are then compared with a non-forecast approach, ending with a discussion and conclusion.

10.2 DESCRIPTION OF APPLICATION SITE

Ethiopia possess abundant water resources and hydropower potential, second only to the Democratic Republic of Congo in all of Africa, yet only two percent of this potential has been developed (World Energy Council, 2007). Currently, 83% of Ethiopia's population lacks access to electricity, with 94% still relying on fuel wood for daily cooking and heating (Tegenu, 2006). The Ethiopian government is therefore pursuing ambitious plans and programs to develop hydropower in an effort to substantially reduce poverty and create an atmosphere for social change. It has been shown that access to electricity, including rural electrification, is a key to poverty reduction in Ethiopia (MoFED, 2006). Implementation, however, is not trivial, especially due to the large financing and investment challenges, as well as required institutional capacity.

The Blue Nile headwaters emanate at the outlet of Lake Tana in the Ethiopian highlands, and are joined by many important tributaries, draining 180 000 km² in the central and southwestern Ethiopian highlands (Steenhuis et al., 2009) becoming a mighty river long before it reaches the lowlands and crosses into Sudan (Fig. 1). It stretches nearly 850 km between Lake Tana and the Sudan-Ethiopia border, with a fall of 1300 m; the grades are steeper in the plateau region, and flatter along the low lands. Very few stream gauges exist along the Blue Nile River within Ethiopia, and those that do tend to have spotty or limited records, and are often not publicly available. Roseires dam in Sudan presents the first streamflow record of sufficient length; monthly averages are illustrated in Fig. 2.

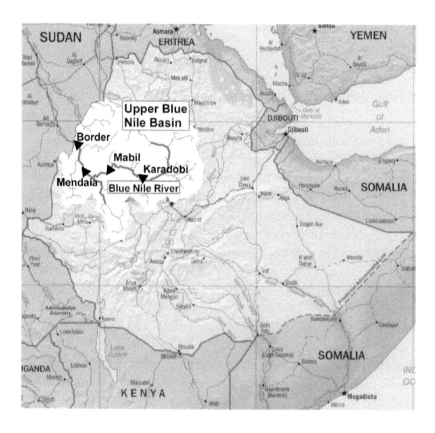

FIGURE 1: The upper Blue Nile basin, Ethiopia, including proposed large-scale hydropower dams. Base map courtesy of the PerryCastaneda Library map collection, University of Texas. ˜

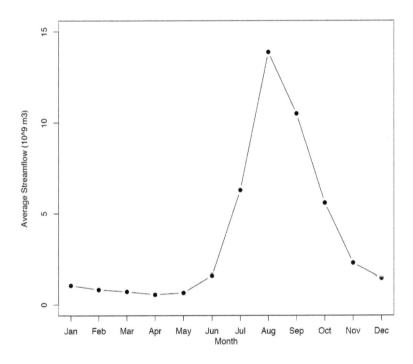

FIGURE 2: Mean monthly streamflow at Roseires, Sudan, 1961–1990. Same as Fig. 1 in Block and Strzepek (2010).

The climate in the Blue Nile River basin varies between its inception in the wet, moderate highlands of Ethiopia and its confluence with the White Nile River in a drier, warmer region. Monthly precipitation records indicate a summer monsoon season, with highest totals in the June–September months (Block and Rajagopalan, 2007); seasonal precipitation averages in excess of 1000 mm in the highlands but only 500 mm near the Sudan border (Shahin, 1985; Sutcliffe and Parks, 1999), with significant interannual variability throughout as illustrated in Fig. 3a (solid line). Near the border, rains during this season account for nearly 90% of total annual precipitation, while in the highlands, approximately 75% of the annual precipitation falls during the monsoon season. The El Nino – Southern Oscillation (ENSO) phenomenon is a main driver of the interannual variability in seasonal precipitation in the basin, with El Nino (La Nina) events generally producing drier (wetter) than normal conditions (Block and Rajagopalan, 2007). Evaporation in the basin varies inversely with precipitation, favoring lesser annual rates in the highlands (~1150 mm) compared with excessive rates (~2500 mm) near the Sudan-Ethiopia border (Shahin, 1985; Sutcliffe and Parks, 1999).

In 1964, the United States Bureau of Reclamation (USBR), upon invitation from the Ethiopian government, performed a thorough investigation and study of the hydrology of the upper Blue Nile basin. Included in the USBR's study was an optimistic list of potential projects within Ethiopia, including preliminary designs of dams for irrigation and hydroelectric power along the main Blue Nile stem. The four major hydroelectric dams along the Blue Nile, as proposed by the USBR, are presented in Fig. 1. Operating in series, these four dams could impound a total of 73 billion cubic meters, which is equivalent to approximately 1.5 times the average annual runoff in the basin. The total installed capacity at design head would be 5570 megawatts (MW) of power, about 2.5 times the potential of the Aswan High Dam in Egypt, and capable of providing electricity to millions of homes. This would be an impressive upgrade over the existing 529 MW of hydroelectric power within Ethiopia as of 2001 (Thomson Gale, 2006). Initial construction costs range from $1.8–$2.2 billion per dam; annual costs (operation and maintenance, scheduled replacement, and insurance) begin in the first year postconstruction and range from $12.5–$17.9 million (Bureau of Reclamation, 1964). While none of these dams have actually been constructed, chiefly due to financial constraints, the Ethiopian Government still has intentions for their full development. For demonstration purposes, however, the dams are assumed online and functional, which is analogous to an operationallevel planning study, ideally providing insight into additional expected benefits with forecast inclusion.

10.3 LINKED MODELING SYSTEM

To evaluate the expected benefits of forecast inclusion, in comparison to climatology-based operations, a linked modeling system approach is adopted. This allows processing and transformation of the Kiremt (June–September) monthly precipitation into streamflow for hydropower optimization along the Blue Nile River. The framework is structured by linking previously developed, independent models.

10.3.1 STRUCTURE AND COMPONENTS

Three major modeling components are required: precipitation forecast, rainfall-runoff, and hydropower/water systems optimization (Fig. 4). The forecast model (Block and Rajagopalan, 2007) predicts total seasonal (June–September) precipitation over the Blue Nile basin. One-season lead (March–May) predictors include sea level pressures, sea surface temperatures, geopotential height, air temperature, and the Palmer Drought Severity Index (PDSI), identified through correlation mapping with seasonal precipitation (e.g. Singhrattna et al., 2005; Grantz et al., 2005). The correlation patterns in sea surface temperatures and sea level pressures resemble ENSO features, yet are more skillful than common ENSO indices (Block and Rajagopalan, 2007). The remaining three predictors capture regional characteristics, with PDSI acting as a soil moisture surrogate.

A nonparametric forecast model based on local polynomial regression (Loader, 1999) is adopted to address various shortcomings common with linear regression, including artificially high skill scores stemming from limited data length and multicollinearity among predictors, regression coefficients being greatly influenced by a small number of outliers, often leading to a poor fit, and the inability to capture nonlinear relationships. In the nonparametric approach, estimation of the model function is performed "locally" at the point to be estimated; therefore no global mathematical relationship between predictors and seasonal precipitation exists. This "local" estimation provides the ability to capture features (i.e. nonlinearities) that might be present locally, without granting outliers any undue influence in the overall fit. Optimal model parameters and predictors (the five previously mentioned) are selected via the generalized cross validation score function (Craven and Wahba, 1979). A detailed implementation algorithm is available in Block and Rajagopalan (2007).

Unique forecast distributions for each year are created by adding normal random deviates (mean zero and standard deviation of the global predictive error) to the forecasted precipitation value (Helsel and Hirsch, 1995). These normal distributions represent the inherent uncertainty of the forecast prediction coming from the statistical model. Figure 3 illustrates the observed and modeled time-series for 1961–2000; forecast distributions are represented as box plots in Fig. 3b. Median values from the forecast distributions demonstrate a correlation coefficient of 0.69 with observations; a rank probability skill score of 0.39, using the full distribution, is a marked improvement over climatology alone. Seasonal precipitation forecasts are disaggregated into monthly forecasts through a proportion vector calibrated on observed data.

The rainfall-runoff model employed is a derivative of the Watbal model (Yates, 1996; Yates and Strzepek, 1998), specifically calibrated to the Blue Nile basin. It is a semidistributed, average-monthly model, with lumped soil and vegetation type and distributed climatic inputs, applied to gridded data ($0.5° \times 0.5°$ for this study). The model simulates changes in soil moisture and runoff, and is essentially an accounting scheme based on a conceptualized, onedimensional bucket that lumps both the root and upper soil layer. The model comprises two elements: the first is a water balance component that describes water movement into and out of a conceptualized basin; the second is the calculation of potential evapotranspiration, which is computed using the FAO Penman-Monteith approach (FAO, 1998). The water balance component of the model comprises three parameters: surface runoff, sub-surface runoff, and maximum catchment water-holding capacity. The simplified representation of soil moisture dynamics has been shown to adequately represent runoff changes due to climate fluctuations (Yates, 1996; Yates and Strzepek, 1998). A final module translates runoff into Blue Nile River streamflow for critical points throughout the basin by aggregating runoff from upstream grid cells, as determined with digital elevation maps.

Given the minimal availability of relevant hydrology data within the basin, rainfall-runoff model calibration and validation is contingent on a long streamflow record at the Roseires dam, just over the border in Sudan. Inflow into potential reservoirs is inferred; no validation at these points is possible. Clearly, additional data collection and sampling for improved rainfall-runoff modeling would be necessary prior to final design of the dams and reservoirs.

The hydropower/reservoir model selected is IMPEND, the Investment Model for Planning Ethiopian Nile Development (Block and Strzepek, 2010). It is classified as a planning tool with operational-level detail to help define feasibility and expectations of project choice. IMPEND is a deterministic water resources system model requiring a single input file of monthly streamflow and net evaporation at the four proposed Ethiopian dam locations and at the existing Roseires dam in southeastern Sudan (all in series). The model thus encompasses the Blue Nile River from its inception at Lake Tana to the Roseires dam, just beyond the Sudan-Ethiopian border. The current version values hydropower at 8-cents per kilowatt-hour; reservoir head represents the decision variable and net present benefits constitutes the objective value.

When linked with the forecast and rainfall-runoff models, IMPEND is analogous to an implicit stochastic optimization process, with two notable exceptions (Loucks et al., 1981; Draper, 2001). First, model foresight is limited to 12 months (not the full time-sequence of analysis), and second, that foresight is imperfect when based on forecasts or climatology (see Sect. 3.3). The model maximizes hydropower benefits given the hydrologic state variables, including the existing reservoir storage volume and limited foresight inflow forecast. Specific model equations and details are provided in Block and Strzepek (2010).

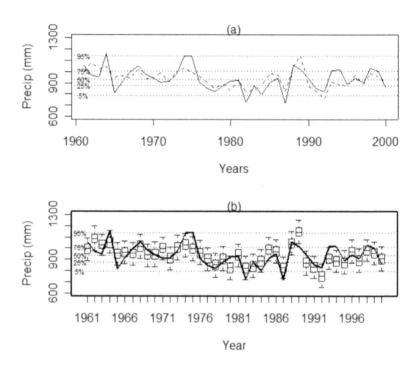

FIGURE 3: Local polynomial June–September precipitation forecast modeling approach results. (a) Observed and cross-validated estimates with horizontal lines at percentiles from the observed seasonal precipitation. (b) Box plots of cross-validated ensembles with horizontal lines at percentiles from the observed seasonal precipitation. Observed data shown as solid line; cross-validated model estimates shown as dashed line in (a) and boxes in (b). Same as Fig. 11 in Block and Rajagopalan (2007).

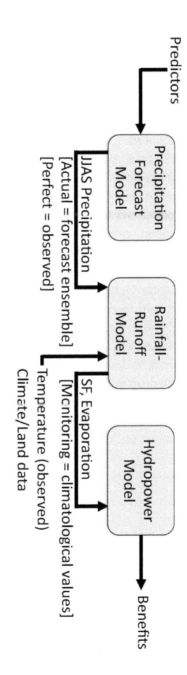

FIGURE 4: The linked modeling system with a monthly time-step. JJAS = June–September, SF = streamflow.

A key attribute of the model is its ability to accept monthly input data varying from year to year, which is critical for proper performance assessment. Analysis based solely on a climatological perspective may well misjudge long-term project benefits (Block et al., 2008). IMPEND is also capable of assessment over various interest (or discount) rates; for the purposes of this study, this rate has been restricted to 10%. This social rate of discounting has been used by others (e.g. Jabbar et al., 2000) and falls within the range of discount rates experienced by Ethiopia within the last five years (Central Intelligence Agency, 2006). A final noteworthy characteristic is the flexibility of downstream flow policies, modulated by the downstream flow constraint established at the entrance to Roseires dam. The policy employed here allows for up to five percent of the annual flow passing the border to be impounded within Ethiopia.

IMPEND may be configured to account for transient (filling) conditions for dams coming online, however for this analysis, all reservoirs are assumed initially filled to design height to mimic operational aspects. Reservoir operations are not restricted to explicitly following established rule curves, but rather are flexible, optimizing operations based on expected future streamflow. Operations are, however, constrained to meet or exceed a future target storage level, as described in Sect. 3.3. Benefits attributable to climate forecasts in hydropower optimization are an aspect not often considered. For a climatology-based operation, however, for which future streamflows are simply based on observed monthly means, this approach essentially collapses to a process analogous to following a rule curve.

10.3.2 DATA

For the forecast model, global atmospheric and oceanic variables, including sea surface temperature, sea level pressure, geopotential height, and air temperature, were obtained from the National Oceanic and Atmospheric Administration's (NOAA's) Climate Diagnostics Center (CDC), based on National Centers for Environmental Prediction–National Center for Atmospheric Research (NCEP-NCAR) reanalysis data (Kalnay et al., 1996). These are monthly average values on a 2.5° × 2.5° grid for 1949 to the present. PDSI values (Dai et al., 2004), also at monthly time scales and on a 2.5∘ × 2.5∘ grid, for 1870–2003, were provided by NCAR's Climate and Global Dynamics Division. Observed precipitation data (the predictand) are part of the Climatic Research Unit (CRU) time series 2.0 dataset, on a 0.5∘ × 0.5∘ grid resolution, obtained from the University of East Anglia, Norwich, United Kingdom (Mitchell and Jones, 2005).

In addition to the June–September monthly precipitation produced by the forecast model, the rainfall-runoff model requires inputs of mean daily temperature and the diurnal temperature range, acquired from the CRU dataset (Mitchell and Jones, 2005). Precipitation for months other than June– September represents climatology based on the same dataset. Monthly streamflow and net evaporation outputs are produced monthly.

Physical, hydrologic, and climatic data required for building and running IMPEND were acquired from a number of sources. Dam, reservoir, and power characteristics are provided in the USBR preliminary study (Bureau of Reclamation, 1964). Observed streamflow records for calibration and validation in the upper Blue Nile basin are publicly available through numerous sources, including NCAR's ds552.1 dataset (Bodo, 2001). The rainfall-runoff model provides streamflow and net evaporation. IMPEND outputs include project benefits or net benefits (discounted back to the simulation start year), energy production, and reservoir levels at monthly intervals.

10.3.3 FORECASTING APPROACHES

To quantify the value of a seasonal precipitation forecast carried through the linked system, three types of forecasts are evaluated through a hindcast: perfect, actual (imperfect), and monitoring (based on climatology). Decision-making is contingent on current conditions (e.g. reservoir storage) and an outlook of precipitation over the ensuing 12 months. In the perfect forecast case, the 12-month foresight period consists of observed precipitation, and represents an upper bound on expected benefits possible through forecast improvement. The actual and monitoring forecast approaches do not use perfect precipitation foresight, but rather a combination of imperfect forecasts and climatology out 12 months, as detailed below. The selection of 12 months, as opposed to a shorter foresight, is based on the large reservoir storage capacity, and also facilitates constraining reservoir levels to prevent dumping for short-term gains in the optimization.

For the perfect forecast case, observed precipitation and temperature are fed into the rainfall-runoff model to generate a streamflow and net evaporation sequence to drive the hydropower model. Given the lack of observational data at the inlet of the reservoirs, the streamflow and net evaporation produced by the rainfall-runoff model based on observed precipitation are substituted for observations. Operational decisions in the hydropower model are optimized in the current step reflective of the existing reservoir storage and inflows over the next 12 months, subject to constraints, including the target storage level. Marching forward one month, the hydropower model receives new information regarding the time-step 12 months out, and reoptimizes reservoir decisions accordingly to maximize benefits. This iterative approach forms a set of sequential deterministic problems (e.g. Stedinger et al., 1984; Marino and Loaiciga, 1985a,b).

The actual forecast approach follows a similar overall progression, excepting application of forecasted precipitation for June–September, with climatological precipitation in other months, as inputs into the rainfall-runoff model. Observed temperatures are also applied. In any given month, the hydropower model optimizes based on existing reservoir storage and inflows over the next 12 months, as in the perfect forecast case, however prior to May, when the forecast is issued, streamflow and net evaporation values are based only on climatological precipitation and persistence. Thus in May of every year, streamflow and net evaporation for each of the subsequent 12 months of the sequence are updated, reflecting the new precipitation forecast.

Decision-making for the monitoring approach is simply based on optimizing over current reservoir conditions and climatological streamflow and net evaporation, the common approach for most water managers.

Thus, for the actual and monitoring approaches, an optimal release is selected in the current month time-step, conditioned on future imperfect forecast or climatological inflows, respectively, and start-of-month reservoir storage. Reservoir storage after releases is subsequently computed via a mass balance using the forecast or climatological inflow. The endof-month reservoir storage, however, must be updated (or corrected) by the difference between the imperfect or climatological inflow and the observed inflow; reservoir storage is increased or decreased accordingly. This is suboptimal, from the hydropower model perspective, but realistic given inflow forecast errors. Finally, the end-of-month reservoir storage becomes the existing storage for the ensuing month, and the iterative optimization scheme is updated (Marino and Loaiciga, 1985a,b). Observed inflows are produced by the rainfall-runoff model utilizing observed precipitation and temperature, as discussed previously.

10.3.4 HYDROCLIMATIC SEQUENCES DEVELOPMENT AND EVALUATION

To force the linked modeling system, decadal sequences of monthly hydroclimatic input variables (precipitation forecast, temperature) need to be assembled. It is commonly accepted that forecast distributions better capture observations in some years (or seasons) than others, thus the choice of decades allows for examination of compounding effects and tends to smooth out noisy monthly optimization behavior. The goal here is to compare adoption of approaches over time, not for any isolated event. To start, four decadal sequences are constructed from the model hindcast period (1961–1970, 1971–1980, 1981–1990, 1991–2000), providing a comparison over varying climatic conditions. The perfect approach applies observed precipitation, the actual approach applies means of the imperfect precipitation forecast distribution, and the monitoring approach applies climatological precipitation. To capture forecast uncertainty in the actual approach and translate it into benefits, 100 additional sequences are assembled for each of the four decades by randomly drawing from the forecast distribution of the appropriate year (Datta and Houck, 1984; Datta and Burges, 1984). Each of the 100 sequences is then processed through the system of models, including optimizing operations contingent on inflows, existing reservoir storage, and a target storage level.

The number of sequences may be further augmented by bootstrapping from the observed record and in the case of the actual approach, drawing from the forecast distributions. The goal of these additional sequences is to better understand the plausible effects of potential climatic variability (e.g. juxtaposition of wet or dry periods not experienced in the observed record) and model uncertainty (including forecast and hydrology). This is analogous to a sensitivity-type approach to appreciate potential effects on system benefits.

One-hundred bootstrapped sequences are formed for all three approaches simultaneously and processed through the linked model system according to the following algorithm (see Fig. 4 for flow chart):

1. Randomly select a year, 1961–2000.
2. *Perfect:* Retain monthly precipitation observations.
   Actual: Draw from the precipitation forecast distribution from the year selected (Fig. 3) and disaggregate to June–September months. For each non June–September month, fill the precipitation record with monthly climatological values.
3. Repeat steps (1)–(2) ten times (with year replacement) to form a decade-long record.
4. Perfect and Actual: Process the sequence through the rainfall-run-off model to produce monthly net evaporation and streamflow into the reservoir.
5. *Perfect and Actual:* process the net evaporation and streamflow sequences through the IMPEND hydropower model to generate hydropower benefits. Actual approach requires reservoir storage update based on observed inflows.
   Monitoring: process decade-long climatological net evaporation and streamflow sequences through IMPEND to generate hydropower benefits. Requires reservoir storage update based on observed inflows.
6. Repeat steps (1)–(5) one hundred times.

An evaluation of the temporal persistence of interannual June–September total seasonal precipitation reveals no particular persistence at any lag, therefore interannual stream- flow may be considered a random process, justifying a bootstrap approach to construct decadal sequences. Monthly persistence between years, specifically December to January, exists weakly, but has not been preserved for the perfect approach; this is the dry season, and variability (let alone quantity) is small, producing negligible differences. For the actual and monitoring approaches, climatological values are assumed for December and January, so persistence is irrelevant.

The three forecasting approaches are evaluated by comparing the sum of monthly hydropower benefits, aggregated to decadal totals. Initial comparisons include approaches with all four proposed dams online; latter comparisons include only Karadobi, the farthest upstream dam site.

TABLE 1: Decadal reliability and annual resilience for two forecasting approaches in comparison to monitoring (climatology). Four-reservoir scheme only, with sequences based on observed decades in chronological order.

Forecast Approach	Decades							
	1960s		1970s		1980s		1990s	
	Rel	Res	Rel	Res	Rel	Res	Rel	Res
(Four-reservoir scheme)								
Full forecast	0	0.33	0.29	0.37	0.97	0.52	0.84	0.51
BN tercile forecast only	0.13	0.70	1.0	0.60	1.0	0.58	0.85	0.48

Note: BN = below normal precipitation; Rel = decadal reliability, Res = annual resilience

Two performance metrics, analogous to reliability and resilience, are created for further comparison between actual forecast and monitoring benefits. Reliability is represented as:

$$\text{If } FB_t > MB_t \ z_t = 1, \text{ else } z_t = 0 \tag{1}$$

$$\text{Reliability} = (\Sigma_t z)/n \tag{2}$$

where FBt represents hydropower benefits from the actual forecast model system at time-step t, MB are the monitoring approach system benefits, z is a counting scalar, and n equals the total number of time-steps. This comparative reliability may therefore vary from 0–1; values less than 0.5 infer higher overall reliability by the monitoring methodology, while values greater than 0.5 indicate higher overall reliability by the actual forecast approach. Resilience measures the ability of the actual forecast system to respond to years with benefits lower than the monitoring approach with greater benefits in the following year.

If $FB_t < MB_t$ and if $FB_{t+1} > MB_{t+1}$ then $y_t = 1$ (3)

If $FB_t < MB_t$ and if $FB_{t+1} < MB_{t+1}$ then $y_t = 0$ (4)

Resilience $= (\Sigma_{t-1}\ y)/m$ (5)

where y is a counting scalar, and m equals the total number of occurrences. Resilience varies from 0–1 with larger values signifying greater actual forecast system resilience.

10.4 ECONOMIC VALUE AND RELIABILITY OF SEASONAL CLIMATE FORECASTS

The benefits of the linked model system drawing on the three forecasting approaches, reliabilities between the actual forecasting and monitoring approaches, and application of a tailored actual forecast to address water manager's riskaversion are presented in the following section.

10.4.1 CHRONOLOGICAL ANALYSIS OF OBSERVED RECORD

Using the four decades from the observed record, decadal hydropower benefits of the linked model system from the perfect, actual forecast, and monitoring approaches are assessed (Fig. 5). Box plots characterize the expected benefit distribution of the 100 sequences drawn from the forecast distributions. Dams are assumed to be online from the onset. Only the 4-reservoir scheme is displayed, as the single reservoir (Karadobi only) scheme behaves quite similarly. As expected, the perfect forecast outperforms both the actual and monitoring forecasts; with the exception of the first decade, the actual forecast system benefits are on par or surpass those of the monitoring system. Poor forecasts in the early years, especially 1962–1963, for which notably wetter than observed conditions are predicted, contribute to the slightly inferior actual forecast system performance of that decade. The third decade, for which the actual forecast approach benefits far exceed those of the monitoring approach, is a relatively dry period with two exceptionally dry years (1982, 1987). The actual forecast model does predict drier than normal conditions for those years, but not to the extreme observed. In summary, while for three decades there is little appreciable difference between the median-based forecast and the no-forecast approaches, for the 1980s decade, the difference is stark; it is these major hits that water managers wish to avoid. The range of actual forecast decadal benefits, representing forecast uncertainty, is not trivial, as illustrated by the box plots, especially for the first three decades. For the 1990s decade, observations are predominantly in the interquartile range, with limited extreme years, thus producing a very tight benefit distribution. Not surprisingly, benefits derived from the median forecasts are superior to the median of decadal sequences based on randomly drawing from the forecast distributions. Reliability and resilience of the actual approach based on the 100 sequences, as compared with the monitoring approach, are very low for the first two decades, and starkly higher for the last two (Table 1). From this preliminary analysis, it would appear there is value in using an actual seasonal forecast, compared with climatology, to potentially reduce variability in benefits and buffer against considerable losses, however the low reliability and resilience in the first decades is concerning.

10.4.2 SAMPLING FROM THE OBSERVED RECORD

A comparison of decadal benefits between the actual forecast and monitoring system approaches for the 100 bootstrapped sequences, under both the 4-reservoir and single-reservoir (Karadobi only) schemes, is illustrated in Fig. 6. Points above the 1-to-1 line represent sequences for which the actual forecast method's cumulative benefits surpass those of the monitoring method; similarly, points below the line favor the monitoring method. For both the multi and single reservoir schemes, the majority of points are bundled around the 1-to-1 line at the higher benefit end, not clearly favoring either forecasting approach. For sequences resulting in lower benefits, the actual forecast method tends to fare better. Negative benefits are possible due to a penalty function applied to low reservoir levels/low outlet flows, but contribute minimally. Reliability for the single reservoir scheme is 0.68 and 0.58 for decadal and annual (1000 years) sequences, respectively, indicating added value for actual forecast inclusion (Table 2). Resilience is 0.48 for the annual assessment, indicating an immediate rebound for approximately half the occurrences (Table 2). Lack of resilience may be explained by multiple sequential poor forecasts, or the inability of the system to respond timely to a poor forecast, even if the subsequent year's forecast is adequate. The multireservoir scheme reliability and resilience (not reported) is almost identical.

Although the overall results are generally positive, specific sequences, such as decades labeled A and B on Fig. 6, may be severe enough to dissuade managers from accepting an actual forecasting approach. Even though the likelihood of these events occurring is small, the risk may still be suffi- cient. Examining these specific sequences in detail is enlightening. Annual streamflow and benefits for decade A from the actual and monitoring forecast approaches, for the single-reservoir scheme, are illustrated in Fig. 7. Most notably from the streamflow sequence is the over-prediction by the actual forecast system in years 2–3 and 5. The ramifications of this are evident in the annual benefits figure: in years following a poor forecast, benefits drop noticeably in comparison to the monitoring approach, especially when forecasting greater than observed "wet" conditions. (The figure illustrates discounted benefits, so a general downward trend is not unexpected.) This phenomenon is also apparent upon inspection of decade B, presented in Fig. 8, in which years 3 and 5–7 all represent forecasts greater than observed. Similar findings explain the poor performance of the actual forecast for decades A and B under the multi-reservoir scheme (not shown).

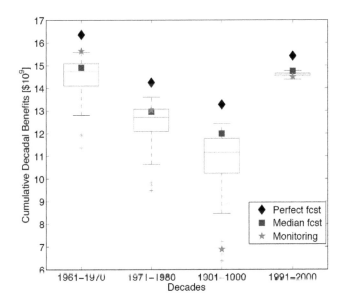

FIGURE 5: Cumulative decadal hydropower benefits for four observed decades using the perfect (diamond), actual (median, square), and monitoring (climatology, star) precipitation forecasts. Box plots characterize the expected benefit distribution of the 100 sequences drawn from the forecast distributions.

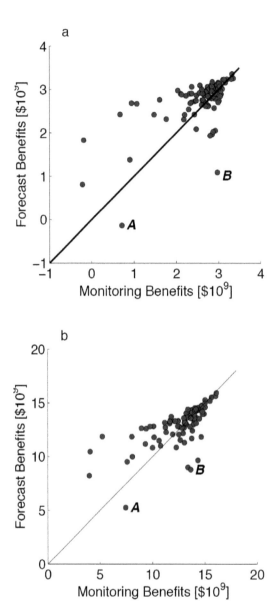

FIGURE 6: Comparison of hydropower benefits between monitoring and actual forecast linked model approaches for (a) single-reservoir (Karadobi), and (b) 4-reservoir schemes.

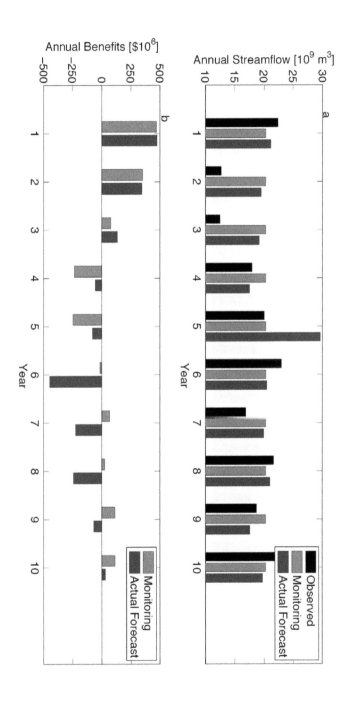

FIGURE 7: Annual analysis of decadal time-series A as identified in Fig. 6. (a) Annual streamflow based on observed, monitoring (climatology), and actual forecast precipitation. (b) Annual discounted hydropower benefits based on monitoring and actual forecast linked model approaches.

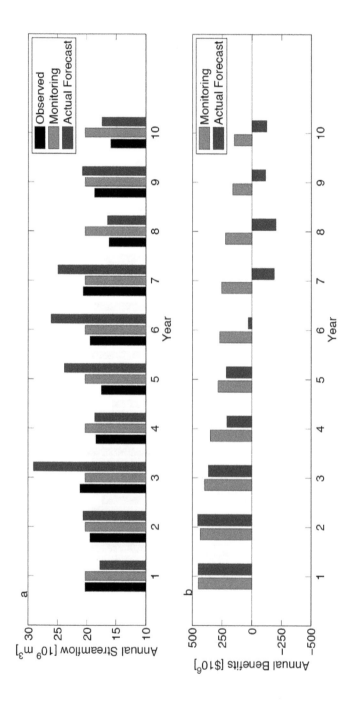

FIGURE 8: Same as Fig. 7, except using time-series B from Fig. 6

TABLE 2: Decadal and annual reliability and resilience for two forecasting approaches in comparison to monitoring (climatology). Single-reservoir scheme only based on bootstrapped sequences.

Forecast Approach		Reliability	Resilience
(single reservoir scheme)	Decadal	Annual	Annual
Full forecast	0.68	0.58	0.48
BN tercile forecast only	0.97	0.72	0.69

Note: BN = below normal precipitation

10.4.3 TAILORING THE SEASONAL FORECAST

Undeniably, water managers considering implementation of a seasonal forecast into operations would prefer the vast majority (or all!) of the project benefits reside above the 1-to-1 line. While the results of the actual forecast system demonstrated thus far indicate potentially greater benefit value versus a monitoring approach, tailoring the forecast in response to hydropower model sensitivities is worth exploring.

Examining the actual forecast errors (1961–2000) by terciles (i.e. below normal, near normal, above normal) reveals approximately equal error means and standard deviations across each category, implying no inherent bias in the forecast. Hydropower response to forecast errors across terciles, however, is less even; the model appears to express much greater sensitivity to forecast errors in the above normal category versus the below normal one. To test, a synthetic sequence is created for the above (below) normal category: June–September precipitation for each year is set at the 75th percentile (25th percentile) of the observational record to represent consistently above (below) normal conditions; remaining months are set to climatology. To mimic a forecast error, two additional sequences are created by adding (subtracting) 25 mm of monthly precipitation from each of the June–September months to the above (below) normal synthetic sequence. The choice of 25 mm is arbitrary. Both sequences are passed through the linked model system (no forecast). The synthetic sequences act as "observations" (constant over the decade for simplicity) for comparison with sequences containing forecast errors through a systematic bias; this allows for gauging the effects of forecast errors in the two terciles in a fair manner. Comparison of differences between "observed" and "observed with forecast error" sequences (Table 3) clearly exhibits greater hydropower model sensitivity to above normal forecast errors. Thus, for this linked model system, errantly forecasting wetter than observed conditions in the above normal category appears to be more detrimental to hydropower operations and ensuing benefits than errantly forecasting drier than observed conditions in the below normal category. This typically stems from aggressive actions (e.g. significant release from storage) following a wet forecast as opposed to conservative actions (e.g. maintaining storage) following a dry forecast. Therefore dampening "wet" forecasts and retaining "dry" forecasts is a reasonable option in light of this sensitivity.

TABLE 3: Results of hydropower sensitivity test to forecast errors in above and below normal terciles. Single-reservoir scheme only based on bootstrapped sequences.

Series	Aggregated Benefits (million $)		Difference
(single reservoir scheme)	Observed	Observed with Error	
AN series (75th percentile)	3682	2272	1410, 38.3%
BN series (25th percentile)	2975	2745	230, 7.8%

Notes: AN = above normal precipitation, BN = below normal precipitation; percentile based on observational 1961–2000 record; error is +25 mm/mo for June–September for AN, −25mm/mo for BN

To this end, the precipitation forecast is tailored such that all actual forecasts in the above normal and near normal terciles are modulated to reflect climatology. (Little is gained from a near normal forecast in comparison to climatology.) Practically, this implies replacing occurrences of June–September forecasts of near or above normal conditions with June–September average monthly precipitation. This procedure effectively eliminates wet forecasts; some opportunities are clearly lost, however damages due to poor wet forecasts deem them worthy of disregarding. Actual forecasts in the below normal category remain unchanged. This modification has no effect on the perfect or monitoring forecast system approaches. The effect of this tailored forecast in comparison to the monitoring approach is quite dramatic (Figs. 9 and 10, updating Figs. 5 and 6a, respectively). Decadal and annual reliability and resilience scores, presented in Tables 1 and 2, indicate a marked improvement over the actual full forecast approach. For the first three observed decades (Fig. 9), the benefit distributions are notably higher and tighter; box plot medians surpass the monitoring bene- fits for all decades except the 1960s. Even though reliability for the 1960s is still relatively low, the difference between benefits from the actual and monitoring approaches has been reduced substantially. Decadal sequence benefits from the bootstrapped actual tailored forecast approach nearly always outpace those of the monitoring approach, with the few favoring the monitoring approach in close proximity to the 1- to-1 line. The elimination of low or negative decadal benefits from the actual forecast system is promising, and may begin to entice managers to incorporate such methodologies into their practices. The relatively low annual reliability reported for the bootstrapped sequences, may be deceiving, and is best understood in context. To take an example, in a dry year, benefits from the monitoring approach may outpace the actual forecast approach, as it prescribes the release of more water through the turbines that year, however repercussions to benefits in the following year are likely to be more severe for the monitoring case. The resilience metric addresses this issue, indicating a rebound by the actual forecast approach in the following year for more than two-thirds of the occurrences.

While tailoring the actual forecast to this stage is clearly beneficial, incentives to improve the forecast model to potentially draw even greater returns is evident through comparison with the perfect forecast output (Fig. 11, bootstrapped sequences only). Attaining a perfect forecast may be unrealistic due to inherent climate uncertainty, however the potential for further advancement plainly exists.

10.4.4 ASSESSING BEHAVIORAL RISK OUTCOMES

The level of risk a water manager is willing to accept is intrinsically tied to institutional requirements, user demands, the flexibility of the system, and personal experiences, among other influences. This level implies consequent effects on system reliability and benefits. Two tendencies are addressed here through the use of a penalty function: one toward risktaking (RT), one toward risk-aversion (RA). The simple linear functions adopted to impose a penalty (represented by energy loss but effectively financial loss) in the event of a predetermined energy threshold not being exceeded is illustrated in Fig. 12. To simulate RA (RT) behavior, a steep (moderate) sloped penalty function is employed. Lowering monthly energy production below the threshold results in larger penalties.

Assessment of the two risk levels for both the monitoring and forecast approaches is undertaken for four minimum energy thresholds, selected to span conceivably acceptable levels of reliability. The identical 100 bootstrapped decadal sequences from the prior evaluation are utilized to illustrate the monthly threshold—reliability and decadal threshold—benefits relationships (Fig. 13). Reliability here in the traditional sense refers to the number of months the threshold is exceeded for the 12 000 months evaluated. Reliability and benefits substantially work in contrast to one another: higher (lower) reliability implies a reduction (increase) in benefits. Also, as the threshold level drops, the difference between levels of risk diminishes, becoming less of a factor when thresholds are easily surpassed. Of notable interest is the clear separation not simply between the monitoring and forecast approaches, but specifically between the RT monitoring and RA forecast. The RA forecast appears more stable, providing greater benefits and higher reliability over the course of thresholds evaluated. Even this conservative behavior produces superior performance when climate information is exploited, perhaps enticing mangers to consider forecast inclusion for improvements in reliability and benefits.

10.5 DISCUSSION AND CONCLUSIONS

The modeling system is necessarily multi-disciplinary, linking climate, hydrology, and water management, an approach to valuing climate information that is often neglected due to its challenges and time consuming nature (Mjelde, 1999). The independent models themselves do not constitute new methodology; the uniqueness of the contribution comes in model integration, the exploitation of sensitivities between integrated models, and ultimately a clear demonstration of economic value through actual forecast inclusion. Ritchie et al. (2004) assert that a forecast system may be considered useful if the forecast is statistically valid (verified) and demonstrates a positive value of information, both of which appear true for this study.

The realization of added value and reliability through forecast inclusion, specifically addressed by dynamic management and decision-making through tailored climate information, is an important outcome. The retention of dry state forecasts adds quality information without subjecting the hydropower model to unreasonable levels of operational risk. Dry forecasts typically prescribe conservative reservoir action, and even if in error, will only forfeit minimal benefits (a higher than expected rainfall will still deliver streamflow to the reservoir for use in later months). Although only exploiting a subset of the forecast range, and likely sacrificing benefits in wet years, a water manager may be inclined to adopt a mechanism that focuses more on reducing risk and potentially lost benefits than lost opportunities. This begins to address one historical impediment to forecast inclusion of not focusing sufficiently on user needs and applications (Ziervogel et al., 2010).

Equally informative is the recognition of forecast benefit from a risk perspective. Risk-averse managers typically face constraints coercing conservative action, whereas risk-takers have more latitude to absorb a low-output time-step in exchange for a substantial payoff later, typically leading toward greater aggregate benefits. Given the success of forecast inclusion demonstrated, it is rather expected that for a specified level of risk, utilizing a forecast produces benefits and reliability in excess of simply depending on climatology. More enlightening is how even conservative action bolstered by a forecast regularly outperforms a risk-taking approach conditioned on climatology, for equivalent energy threshold requirements. This addresses one of the cardinal impediments (risk-aversion) by theoretically allowing managers to remain risk-averse and realize considerable gains.

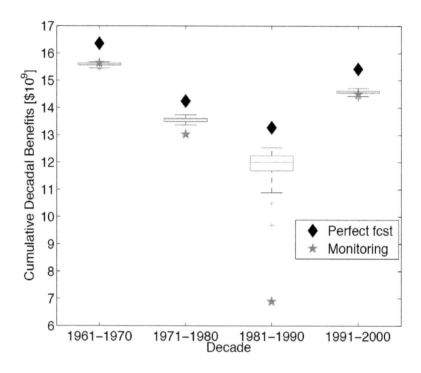

FIGURE 9: Cumulative decadal hydropower benefits for four observed decades using the perfect (diamond), monitoring (climatology, star), and actual (box plots) precipitation forecasts for the four-reservoir scheme. Box plots characterize the expected benefit distribution of the 100 sequences drawn from the forecast distributions tailored to dampen above and near normal precipitation forecasts.

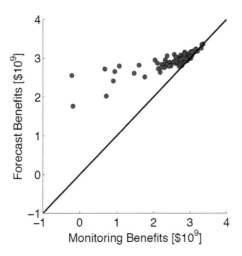

FIGURE 10: Comparison of hydropower benefits between monitoring and actual tailored forecast linked model approaches for the singlereservoir (Karadobi) scheme using the bootstrapped sequences. Tailored approach includes dampening of above and near normal precipitation forecasts.

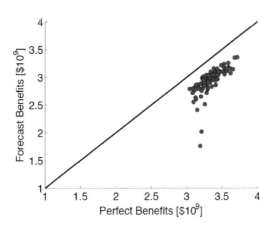

FIGURE 11: Comparison of hydropower benefits between perfect and actual tailored forecast model approaches for the single-reservoir (Karadobi) scheme using the bootstrapped sequences.

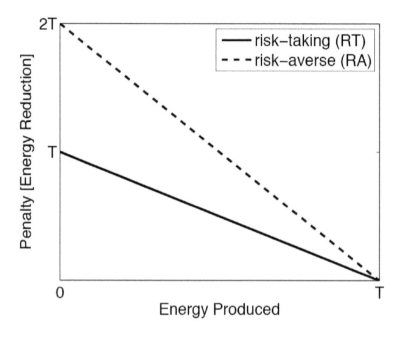

FIGURE 12: Penalty functions for risk-taking (RT) and risk-averse (RA) behaviors. T = energy threshold. Units for Energy Produced and Penalty are GWatt h/mo.

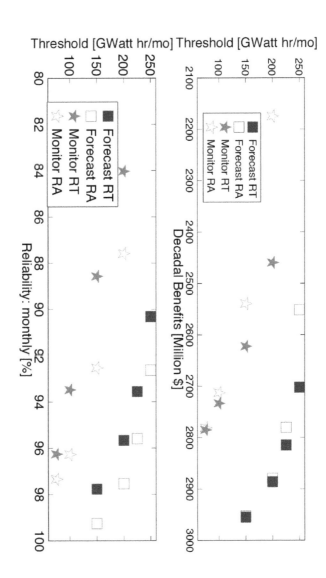

FIGURE 13: Comparison of monitoring and forecast approaches under risk-taking (RT) and risk-averse (RA) behavior for (a) threshold – benefit and (b) threshold – reliability relationships using the bootstrapped sequences.

While the tailored approach demonstrated in this study is effective, it is errant to assume an identical procedure conducted for other regions or project types will necessitate improved benefits, however given success here, exploration is warranted. Locations susceptible to climate variability where forecast skill is apparent are ripe for investigation. As in this instance, it is also conceivable that decisions in other sectors (agriculture and food security, flood early warning, health, etc.) may likewise tend toward emphasis on one end of the forecast spectrum, although minimal evidence of demonstration currently exists.

In addition to climate, sensitivity to coincident drivers (policy, economics, demand, etc.) may also exhibit signifi- cant influence, warranting a reassessment of the hydropower model. Explicitly modeling operational practices, such as selection of a target precipitation or streamflow forecast based on exceedance probabilities (e.g. 90th percentile) of the ensemble forecast, also deserves attention. Other integrated and social factors may also be equally relevant and worthy of inclusion, beyond hydropower simulation, as advocated by the World Commission on Dams (WCD, 2000). The interrelationship of all these aspects is not trivial or currently well understood, however, potentially complicating decisionmaking (McCartney, 2007).

Minimal difference in streamflow was noted when substituting climatological temperature for observed temperature in the rainfall-runoff model. Although the absolute hydropower benefits may be slightly overestimated, from a comparison perspective between forecast and non-forecast approaches, it is essentially immaterial, as both approaches benefit nearly equivalently. Adding a temperature forecast, conditioned on the precipitation forecast, is an added layer of complexity for future stages of this research.

The forecast model presented here is generally founded on stationary climate principles. Multi-decadal effects or climate change are not explicitly addressed. A recent study by O'Connor et al. (2005) reveals the strongest determinant of forecast use is risk perception; if extreme climate conditions are anticipated in the near future, the uptake of a forecast is more likely. Climate change may pose such a perception. Integrating an adaptive forecast model into the linked system to explore implications of changing precipitation trends and variability and increasing temperature warrants future attention.

REFERENCES

1. Axel, J., and Ceron, J. P.: Elements for Life, J. Griffiths, Ed., Tudor ´ Rose, London, 70–71, 2007.
2. Barnston, A. G., van den Dool, H. M., Rodenhuis, D. R., Ropelewski, C. R., Kousky, V. E., O'Lenic, E. A., Livezey, R. E., Zebiak, S. E., Cane, M. A., Barnett, T. P., Graham, N. E., Ji, M., and Leetmaa, A.: Long-Lead Seasonal Forecasts – Where Do We Stand?, B. Am. Meteorol. Soc., 75, 2097–2114, 1994.
3. Barnston, A. G., Kumar, A., Goddard, L., and Hoerling, M. P.: Improving seasonal prediction practices through attribution of climate variability, B. Am. Meteorol. Soc., 86(1), 59–72, 2005.
4. Block, P. and Strzepek, K.: Economic Analysis of Large-scale Upstream River Basin Development on the Blue Nile in Ethiopia Considering Transient Conditions, Climate Variability, and Climate Change, J. Water Res. Pl.-ASCE, 136(2), 156–166, 2010.
5. Block, P. and Rajagopalan, B.: Interannual Variability and Ensemble Forecast of Upper Blue Nile Basin Kiremt Season Precipitation, J. Hydrometeorol., 8(3), 327–343, 2007.
6. Block, P., Strzepek, K., Rosegrant, M., and Diao, X.: Impacts of Considering Climate Variability on Investment Decisions in Ethiopia, J. Agr. Econ., 39, 171–181, 2008.
7. Bodo, B.: Monthly Discharge Data for World Rivers, http://dss. ucar.edu/datasets/ds552.1 (last access: June 2005), 2001.
8. Broad, K., Pfaff, A., Taddei, R., Sankarasubramanian, A., Lall, U., and Souza Filho, F. A.: Climate, stream flow prediction and water management in northeast Brazil: societal trends and forecast value, Climatic Change, 84(2), 217–239, 2007.
9. Bureau of Reclamation: US Department of Interior, Land and Water Resources of Blue Nile Basin, Ethiopia, Main Report and Appendices I–V, Government Printing Office, Washington, DC, 1964.
10. Chiew, F. H. S., Zhou, S. L., and McMahon, T. A.: Use of seasonal streamflow forecasts in water resources management, J. Hydrol., 270(1–2), 135–144, 2003.
11. Central Intelligence Agency – CIA: The World Factbook, CIA, Washington, DC, 2006.
12. Craven, P. and Wahba, G.: Smoothing noisy data with spline functions, J. Numer. Math., 31, 377–403, 1979.
13. Dai, A., Trenberth, K., and Qian, T.: A global data set of Palmer Drought Severity Index for 1870–2002: Relationship with soil moisture and effects of surface warming, J. Hydrometeorol., 5, 1117–1130, 2004.

14. Datta, B. and Burges, S. J.: Short-term, single, multiple purpose reservoir operation: Importance of loss function and forecast errors, Water Resour. Res., 20(9), 1167–1176, 1984.
15. Datta, B. and Houck, M. H.: A stochastic optimization model for real-time operation of reservoirs using uncertain forecasts, Water Resour. Res., 20(8), 1039–1046, 1984.
16. Draper, A.: Implicit Stochastic Optimization with Limited Foresight for Reservoir Systems, Ph.D. Dissertation, University of California – Davis, 2001.
17. Faber, B. A. and Stedinger, J. R.: Reservoir optimization using sampling SDP with ensemble streamflow prediction (ESP) forecasts, J. Hydrol., 249(1–4), 113–133, 2001.
18. FAO – Food and Agriculture Oranization of the UN Crop Evapotranspiration: Guidelines for Computing Crop Water Requirements,
19. FAO Irrigation and Drainage Paper No. 56, Food and Agriculture Organization of the United Nations, Rome, 300 pp., 1998.
20. Goddard, L., Barnston, A. G., and Mason, S. J.: Evaluation of the IRI's "net assessment" seasonal climate forecasts: 1997–2001, B. Am. Meteorol. Soc., 84, 1761–1781, doi:10.1175/BAMS-84- 12-1761, 2003.
21. Grantz, K., Rajagopalan, B., Clark, M., and Zagona, E.: Seasonal Shifts in the North American Monsoon, J. Climate, 20(9), 1923– 1935, 2005.
22. Hamlet, A. F., Huppert, D., and Lettenmaier, D. P.: Economic value of long-lead streamflow forecasts for Columbia River hydropower, J. Water Res. Pl.-ASCE, 128(2), 91–101, 2002.
23. Hellmuth, M. E., Moorhead, A., Thomson, M. C., and Williams, J.: Climate Risk Management in Africa: Learning from Practice, International Research Institute for Climate and Society (IRI), Columbia University, New York, 104 pp., 2007.
24. Helsel, D. and Hirsch, R., Statistical methods in water resources, Elsevier Science, Amsterdam, 1995.
25. Jabbar, M., Pender, J., and Ehui, S.: Policies for sustainable land management in the highlands of Ethiopia: Summary of papers and proceedings of a seminar held at ILRI, Addis Ababa, Ethiopia, 22–23 May 2000, Socio-economics and Policy Research Working Paper 30, ILRI – International Livestock Research Institute, Nairobi, Kenya, 2000.
26. Johnston, P. A., Ziervogel, G., and Matthew, M.: The uptake and usefulness of weather and climate information among water resource managers, Papers Appl. Geog. Conf. 30, 380–389, 2007.
27. Kalnay, E., Kanamitsu, M., Kistler, R., Collins, W., Deaven, D., Gandin, L., Iredell, M., Saha, S., White, G., Woollen, J., Zhu, Y., Chelliah, M., Ebisuzaki, W., Higgins, W., Janowiak, J., Mo, K. C., Ropelewski, C., Wang, J., Leetmaa, A., Reynolds, R., Jenne, R., and Joseph, D.: The NCEP/NCAR reanalysis 40-year project, B. Am. Meteorol. Soc., 77, 437–471, 1996.
28. Karamouz, M. and Vasiliadis, H.: Bayesian stochastic optimization of reservoir operation using uncertain forecasts, Water Resour. Res., 28(5), 1221–1232, 1992.
29. Kim, Y.-O. and Palmer, R.: Value of seasonal flow forecasts in Bayesian stochastic programming, J. Water Res. Pl.-ASCE, 123(6), 327–335, 1997.

30. Lemos, M. C.: What Influences Innovation Adoption by Water Managers? Climate Information Use in Brazil and the United States, J. Am. Water Resour. Assoc., 44(6), 1388–1396, 2008.
31. Loader, C.: Local Regression Likelihood, Springer, New York, 1999.
32. Loucks, D. P., Stedinger, J. R., and Haith, D. A.: Water resources systems planning and analysis, Prentice-Hall, Englewood Cliffs, NJ, 1981.
33. Marino, M. A. and Loaiciga, H. A.: Dynamic model for multireservoir operation, Water Resour. Res., 21(5), 619–630, 1985a.
34. Marino, M. A. and Loaiciga, H. A.: Quadratic model for reservoir management: Application to the Central Valley Project, Water Resour. Res., 21(5), 631–641, 1985b.
35. Maurer, E. P. and Lettenmaier, D. P.: Potential Effects of Long-Lead Hydrologic Predictability on Missouri River Main-Stem Reservoirs, J. Climate, 17, 174–186, 2004.
36. McCartney, M. P.: Decision support systems for large dam planning and operations in Africa, IWMI Working Paper 119, International Water Management Institute, Colombo, Sri Lanka, 47 pp., 2007.
37. Millner, A.: What Is the True Value of Forecasts?, Weather Clim. Soc., 1, 22–37, 2009.
38. Mitchell, T. and Jones, P.: An improved method of constructing a database of monthly climate observations and associated highresolution grids, Int. J. Climatol., 25, 693–712, 2005.
39. Mjelde, J.: Value of Improved Information in Agriculture: Weather and Climate Forecasts, Paper for the Southern Region Information Exchange Group-10 annual meeting, Huntsville, AL, 20–21 May, 32 pp., 1999.
40. MoFED – Ministry of Finance and Economic Development Ethiopia: Status Report on the Brussels Programme of Action for Least Developed Countries, Addis Ababa, Ethiopia, www.un.org/special-rep/ohrlls/ldc/MTR/Ethiopia.pdf (last access: March 2010), 2006.
41. O'Connor, R. E., Yarna, B., Dow, K., Jocoy, C., and Carbone, G.: Feeling at Risk Matters: Water Managers and the Decision to Use Forecasts, Risk Anal., 25(5), 1265–1275, 2005.
42. Pagano, T., Hartmann, H., and Sorooshian, S.: Using climate forecasts for water management: Arizona and the 1997–1998 El Nino, J. Am. Water Resour. Assoc., 37(5), 1139–1153, 2001. ~
43. Patt, A. G., Ogallo, L., and Hellmuth, M.: Learning from ten years of climate outlook forums in Africa, Science, 318, 49–50, 2007.
44. Pulwarty, R. and Redmond, K.: Climate and Salmon Restoration in the Columbia River Basin: The Role and Usability of Seasonal Forecasts, B. Am. Meteorol. Soc., 78(3), 381–397, 1997.
45. Rayner, S., Lach, D., and Ingram, H.: Weather forecasts are for wimps: Why water resource managers do not use climate forecasts, Climatic Change, 69(2–3), 197–227, 2005.
46. Ritchie, J. W., Zammit, C., and Beal, D.: Can seasonal climate forecasting assist in catchment water management decision-making?: A case study of the Border Rivers catchment in Australia, Agr. Ecosyst. Environ., 104(3), 553–565, 2004.

47. Sankarasubramanian, A., Lall, U., Souza Filho, F. A., and Sharma, A.: Improved water allocation utilizing probabilistic climate forecasts: Short-term water contracts in a risk management framework, Water Resour. Res., 45, W11409, doi:10.1029/2009WR007821, 2009.

48. Shahin, M.: Hydrology of the Nile Basin, Elsevier Science, Amsterdam, The Netherlands, 1985.

49. Singhrattna, N., Rajagopalan, B., Clark, M., and Krishna Kumar, K.: Forecasting Thailand summer monsoon rainfall, Int. J. Climatol., 25, 649–664, 2005.

50. Stedinger, J. R., Sule, B. F., and Loucks, D. P.: Stochastic dynamic programming models for reservoir optimization, Water Resour. Res., 20(11), 1499–1505, 1984.

51. Steenhuis, T. S., Collick, A. S., Easton, Z. M., Leggesse, E. S., Bayabil, H. K., White, E. D., Awulachew, S. B., Adgo, E., and Ahmed, A. A.: Predicting discharge and sediment for the Abay (Blue Nile) with a simple model, Hydrol. Process., 23, 3728–3737, 2009.

52. Sutcliffe, J. and Parks, Y.: The hydrology of the Nile, IAHS Press, Wallingford, Oxfordshire, UK, 1999.

53. Tegenu, A.: Statement at the Fourteenth Commission on Sustainable Development, United Nations, New York, www.un.org/esa/sustdev/csd/csd14/statements/ethiopia 11may.pdf (last access: March 2010), 2006.

54. Thomson Gale Encyclopedia of the Nations – Africa: Thomson Corporation, Farmington Hills, MI, USA, 2006.

55. WCD – World Commission on Dams, Dams and Development: A New Framework for Decision-making, Earthscan Publications Ltd., London, 404 pp., 2000.

56. World Energy Council, 2007: Survey of Energy Resources, World Energy Council, London, 586 pp., 2007.

57. Yao, H. and Georgakakos, A.: Assessment of Folsom Lake response to historical and potential future climate scenarios: 2. Reservoir management, J. Hydrol., 249(1–4), 176–196, 2001.

58. Yates, D.: WatBal: An Integrated Water Balance Model for Climate Impact Assessment of River Basin Runoff, Int. J. Water Resour. D., 12(2), 121–139, 1996.

59. Yates, D. and Strzepek, K.: Modeling the Nile basin under climate change, J. Hydrol. Eng., 3, 98–108, 1998.

60. Yeh, W. W.-G., Becker, L., and Zettlemoyer, R.: Worth of inflow forecast for reservoir operation, J. Water Res. Pl.-ASCE, 108(3), 257–269, 1982.

61. Ziervogel, G., Johnston, P., Matthew, M., and Mukheibir, P.: Using climate information for supporting climate change adaptation in water resource management in South Africa, Climatic Change, 103(3–4), 537–544, 2010.

PART V

SEASONAL ENERGY MANAGEMENT

The Identification of Peak Period Impacts When a TMY Weather File Is Used in Building Energy Use Simulation

JAY ZARNIKAU AND SHUANGSHUANG ZHU

11.1 INTRODUCTION

It is not always clear how weather-sensitive energy efficiency measures will perform at the exact hour(s) of the utility's annual summer or winter system peak. Often, building energy use simulation models are used to obtain 8760 hourly impact estimates for the change in load associated with the efficiency measure based on typical meteorological year (TMY) data. TMY data contain actual months of weather data from different past years. Consequently, the TMY year does not coincide with any actual year and thus cannot be matched against actual demand or load data for a utility

The Identification of Peak Period Impacts When a TMY Weather File Is Used in Building Energy Use Simulation. © *Zarnikau J and Zhu S. Open Journal of Energy Efficiency 3,1 (2014). DOI:10.4236/ ojee.2014.31003. This article is licensed under a Creative Commons Attribution 4.0 International License, http://creativecommons.org/licenses/by/4.0/.*

system or market. The hours associated with the most extreme tempera-tures in a TMY file may not necessarily correspond with a peak in demand in a utility system. Other factors, such as day of the week and the hour within the day, may also play a role. The challenge is to determine which of the 8760 hourly values obtained from a building energy use simulation model to select to represent the demand reduction at the time of the util-ity's system peak.

This topic is of great importance to utility system planners. Electric and natural gas utility systems are constructed largely to meet peak de-mand. Thus, the impact or performance of an energy efficiency measure during peak periods is of keen interest. Energy efficiency measures or demand side management (DSM) programs are valued, in part, based upon the generation and transmission costs which they could potentially displace [1] . Thus, the potential for an energy efficiency measure to reduce demand during the system's peak affects the value of measures and programs.

Various utility regulatory commissions in the US provide specific in-structions for utilities and energy efficiency program administrators to fol-low when selecting the hour(s) associated with peak demand impacts. In California's DEER database, "the demand savings due to an energy ef-ficiency measure is calculated as the average reduction in energy use over a defined nine-hour demand period" [2] . These nine hours correspond with 2 pm to 5 pm during 3-day heat waves. The Mid-Atlantic Techni-cal Reference Manual, used in Maryland, Delaware, and DC, states: "The primary way is to estimate peak savings during the most typical peak hour (assumed here to be 5 pm) on days during which system peak demand typically occurs (i.e., the hottest summer weekdays). This is most indica-tive of actual peak benefits." [3] The New York Public Service Commis-sion instructs: "Program Administrators (PAs) should calculate coincident peak demand savings based on the hottest summer non-holiday weekday during the hour ending at 5 pm." [4] Wisconsin's utility regulatory agency requires that peak demand reduction for weather-sensitive efficiency mea-sures be based on average design-day conditions [5] . In Illinois, Colo-rado, New Jersey, and Maine, coincidence factors are used to estimate peak demand reduction based the impacts of a weather-sensitive efficiency measure on annual energy consumption [6] -[9] . For energy efficiency measures which are not weather-sensitive, a number of states find it acceptable to average the expected energy impacts of the measure over a large number of hours within some "peak period". We are unaware of any regulatory authority having adopted a formal probabilistic approach to estimating the impacts of energy efficiency measures upon the peak demand of a utility or market.

A formal probabilistic approach is attractive because the system peak of a utility or a market may not necessarily coincide exactly with the hottest summer day, historical temporal patterns, design-day conditions, or heat waves. For example, an extremely hot summer temperature reading may not necessarily lead to a summer peak, if the extreme temperature occurs on a weekend (when energy use in the commercial or business sector may be lower) or early in the afternoon (before the occurrence of an after-work peak in household energy use). While extreme temperatures may be the most important determinant of system peak demand, various patterns in energy usage (as might be reflected by the time of day and the month of the year), and other factors may play a role as well. A probabilistic approach can be used to quantify how various factors may contribute to the establishment of a peak in system demand.

11.2 PROPOSED APPROACH

Our proposed approach to matching a seasonal peak on a utility system with a TMY data file involves the following steps:

- Establish the number of hours to be included in the set of peak hours to be predicted each year and season (e.g., summer and winter).
- Use a logistic regression model and hourly data for a number of historical years to estimate the relationship between setting a peak hour and a set of explanatory variables, including a temperature variable and dummy variables representing the time-of-day and month-of-year.
- Use the estimated relationships to assign marginal probabilities to changes in the explanatory variables.
- With the estimated relationships, calculate the probability of setting a peak hour based on TMY weather data.
- Find and average the savings (i.e., the difference between a base and change case) from the outputs of a building energy use simulation model that used the same TMY data file which corresponds to the same hours.

Although system planners often use a single hour or 15-minute interval to measure peak demand, predicting a larger set of peak hours tends to be more practical in the first step. Building energy use simulation models have stochastic algorithms. So if a single pair of model runs (i.e., a base case and a change case) is used to calculate hourly savings, the predicted savings may be biased for any single hour. So, either multiple model runs must be used to average the estimated hourly savings, or a broader definition of peak (i.e., peak hours) must be used. Further, in the analysis of the cost-effectiveness of energy efficiency measures and programs, the demand reduction tends to be valued based on the capital cost of a combustion turbine which normally has an expected annual runtime of 10 to 40 hours. Thus an analysis of the cost-effectiveness of energy efficiency measures and programs may benefit from knowledge of the impacts over a set of hours. Finally, estimating the probability of setting a set of peak hours is much easier than estimating a single peak hour or interval per year with a logistic model. For example, if six years of historical data are used and thus $Y = 1$ on only six instances, more advanced techniques would be required in the estimation (e.g., the use of a prior distribution and Bayesian estimation techniques) than those discussed here. For these reasons, a set of 20 peak hours is used in the examples presented here.

Note that the second step ignores many other very important factors that might affect the timing of the peak, including actions by industrial energy consumers and load-serving entities to respond to wholesale market price spikes. The day of the week is also not considered. However, the inclusion of other variables would prevent the application of this approach when only a TMY weather file and a building energy use simulation model are used to calculate the peak demand reduction associated with an efficiency measure. TMY data are pieced-together from recorded weather during numerous previous years to create a typical year with typical fluctuations. Since the TMY data do not represent weather data from any single "real" year, there would be no way of matching "real" energy price data, the day of the week, or other variables to the fabricated weather data.

Marginal probabilities can be obtained by estimating a logistic regression or logit model [11] . Most statistical software packages can convert the results from a logit model into probabilities [12] [13].

In the final step, either a simple average or a probability-weighted average (with the weights based on the probability of the seasonal peak being set in a particular hour in the TMY data file) could be used to estimate peak demand reduction among those hours within the set of peak demand hours.

11.3 AN EXAMPLE DETERMINATION OF PEAK HOURS

An example is illustrated below to further explain the five steps described above. It is applied to the estimation of both summer and winter peak demand reduction associated with various energy efficiency measures.

Total system electrical load or demand in the Electric Reliability Council of Texas (ERCOT) electricity market is used in this example. The ERCOT electricity market is "settled" based on 15-minute intervals. There are 96 intervals in most days. Interval-level data were converted to hourly values to facilitate the estimation and provide a better match of load to hourly temperature data. The top 20 hours of each summer season of each year, Peak Hour, were coded 1, and all other hours were coded as 0. Variables representing the hours ending 16:00, 17:00, and 18:00 were included to capture time-of-day factors affecting electricity use. All hours before 2 pm and after 6 pm were assumed to have zero probability of being within the set of peak hours and were eliminated from the dataset to facilitate estimation. Additionally, two variables representing the month-of-year (July and August) were also included. Because summer peak loads are largely determined by air conditioning usage in Texas, a variable was constructed to represent the ratio between the actual temperature in a central location within the ERCOT market (Austin) for a given interval and the highest temperature reading during the given year (Relative Max Temp).

The resulting model was thus:

$$\text{Logit (Peak Hour)} = f(\text{Relative Max Temp, Hour16, Hour17, Hour18,} \atop \text{July, August}) \qquad (1)$$

This relationship was estimated using R software as a general linear model with a binomial distribution. The estimated coefficients and p-values from the logistic regression are provided in Table 1.

TABLE 1: Logistic regression statistical results.

	Estimate	p-Value
Intercept	−39.3331	<0.0001
RelativeMaxTemp	36.1022	<0.0001
Hour16	1.7570	0.000131
Hour17	1.9924	<0.0001
Hour18	1.5439	0.001012
July	0.9284	0.016848
August	1.7722	<0.0001

As we can see from Table 1, the coefficient estimates are significant at normally-accepted levels of statistical significance, with the possible exception of the dummy variable denoting the impact of the month of July (relative to the omitted months of June and September).

A unit increase in the relative maximum temperature—the ratio between the actual temperature in a central location within the ERCOT market (Austin) for a given interval and the highest temperature reading during the given year—raises the log of the odds of being included among the peak hours by 36.1022 ceteris paribus.

The coefficient estimate of 1.757 on the variable Hour 16 suggests that the log of the odds of the hour between 3:00 pm and 4:00 pm being among the peak hours (versus the 2:00 pm to 3:00 pm period or hour ending 15:00, the time period not explicitly represented in the model with a variable) is 1.757 time higher, holding all other variables constant. The log odds of setting a peak hour between 4:00 pm and 5:00 pm (Hour 17) versus setting a peak hour between 2:00 pm and 3:00 pm is 1.9924 times higher, holding other variables constant. Similarly, the log odds of the hour from 5:00 pm to 6:00 pm (Hour 18) being among the peak hours is 1.5439 times higher, relative to the omitted period and holding all other variables constant.

For the July and August variables, 0.9284 means the log odds of being a peak hour in July versus being a peak hour in June or September are 0.9284 times higher (which is actually a decrease), and the log odds of being a peak hour in August versus being a peak hour in June are 1.7722 times higher, confirming that summer peaks are most likely to occur in August in Texas.

The coefficient estimates expressed in log odds may be converted to odds ratios, by taking anti-logs.

Once the marginal probabilities are estimated, the probability of each hour of the TMY file being included among the set of peak hours can be calculated. As an example, consider an hour (3:00 pm to 4:00 pm, aka the hour ending 16:00) in August, with an hourly temperature of 100°F, and the annual highest annual temperature being 102°F. The estimated log of the odds ratio of being a peak hour versus being outside the set of peak hours:

$$-39.3331 + 36.1022 \times 100/102 + 1.757 + 1.7722 = -0.4096$$

Thus the probability of obtaining a peak hour during that time and under those conditions is

$$\exp(-0.4096)/(1 + \exp(-0.4096)) = 0.4$$

This calculation may be performed automatically with R software.

The 20 hours in the TMY file assigned the highest probability of being within the set of peak hours are identified in Table 2. Our set of 20 peak hours consists of 7 hours in July and 13 hours in August, all falling within the 3:00 pm to 6:00 pm afternoon time period (i.e., the hours ending 16:00, 17:00, and 18:00). Certainly, temperature prominently determines the probability that an hour falls within the set of 20 peak hours. Yet, the TMY hour ending 17:00 on August 5, 2004 earns the third highest probability, despite having a lower temperature (98.06°F) than some hotter hours (e.g., July 28, 1995 at 16:00 and 18:00). This is because the hour ending 17:00 is more likely to set a peak than the hours ending at 16:00 or 18:00. A probabilistic analysis appropriately takes into consideration both the weather and the time-of-day.

Having estimated the probability of each hour being included among the set of 20 peak hours in this section, we next demonstrate how this information may be used to estimate the impact of energy efficiency measures on peak electricity use when a building energy use simulation model is used to estimate hourly energy consumption using TMY data.

TABLE 2: Twenty peak hours with the highest probability of being included among the set of peak hours.

Date in TMY File	Hour Ending	Temperature in Degrees F	maxtemp	Relative-MaxTemp	logodds	Probability
7/28/1995	17:00	102.02	102.02	1	−0.3101	0.42309
8/5/2004	16:00	100.04	102.02	0.980592041	−0.40961	0.400743
8/5/2004	17:00	98.06	102.02	0.961184082	−0.86764	0.295746
7/28/1995	16:00	100.94	102.02	0.98941384	−0.92768	0.283395
8/20/2004	16:00	98.06	102.02	0.961184082	−1.10304	0.249171
7/28/1995	18:00	100.94	102.02	0.98941384	−1.14078	0.242177
8/20/2004	17:00	96.98	102.02	0.950597922	−1.24982	0.222731
7/27/1995	17:00	98.96	102.02	0.970005881	−1.39295	0.198937
8/3/2004	16:00	96.98	102.02	0.950597922	−1.48522	0.18464
8/4/2004	16:00	96.98	102.02	0.950597922	−1.48522	0.18464
8/11/2004	16:00	96.98	102.02	0.950597922	−1.48522	0.18464
8/19/2004	16:00	96.98	102.02	0.950597922	−1.48522	0.18464
8/26/2004	16:00	96.98	102.02	0.950597922	−1.48522	0.18464
8/3/2004	17:00	96.08	102.02	0.941776122	−1.56831	0.172457
8/4/2004	17:00	96.08	102.02	0.941776122	−1.56831	0.172457
8/19/2004	17:00	96.08	102.02	0.941776122	−1.56831	0.172457
8/26/2004	17:00	96.08	102.02	0.941776122	−1.56831	0.172457
7/27/1995	16:00	98.96	102.02	0.970005881	−1.62835	0.164056
7/24/1995	17:00	98.06	102.02	0.961184082	−1.71144	0.152977
7/26/1995	17:00	98.06	102.02	0.961184082	−1.71144	0.152977

11.4 MATCHING THE SELECTED PEAK HOURS TO ENERGY EFFICIENCY SAVINGS PROFILES

To estimate the impact of an energy efficiency measure upon peak demand, we match the hours with the highest probability of being among the set of peak hours to those same hours in the output from a building energy use simulation model that used the same TMY data file. The average of the energy efficiency measure's hourly savings over those 20 hours provides an estimate of the savings associated with the efficiency measure coincident with the summer peak.

Application of this approach to a simulation of the savings associated with ceiling insulation and air infiltration in an electrically-heated home in Austin is presented here. We also examine the savings from two lighting-related energy efficiency measures.

A whole-home simulation was developed using Energy Gauge, a simulation software tool that uses a DOE-2 simulation engine [14] . Prototype home characteristics were selected using available data on the construction, occupancy, and equipment characteristics of Texas homes, as listed in Table 3. The rows labeled "Ceiling Insulation" and "Air Infiltration" state the base and change conditions.

The simulations assumed differently sized HVAC systems for the analysis of the two weather-sensitive efficiency measures:

- Air infiltration: 2.8 ton air conditioning capacity, 3.5 ton heating capacity
- Ceiling Insulation: 4.3 ton air conditioning capacity, 4.8 ton heating capacity

Table 4 compares estimates of the demand reduction of various scenarios associated with our proposed probabilistic approach with some alternative methods:

- Top 2 Hours of All Peak Months. Select the two hours when the peak hour has most-frequently occurred over the last ten years. Examine impacts during those two hours during every summer weekday during four summer months. Average the impacts over the resulting 170 hours—e.g., the hours ending 17:00 and 18:00 during every summer weekday.

- Heat Wave. The TMY weather files are scanned to locate a three-weekday period that has the highest average temperatures during the peak hours.
- Average Over Peak Period. Estimate a measure's average impact between 1 pm and 7 pm on all summer weekdays over four summer months. (510 hours)

TABLE 3: Home characteristics inputs used in simulation model.

Input	Value	Source
Conditioned Area	1915 square feet	Weighted average total conditioned square feet of Texas single family detached Single Family Dwelling (SFD) homes.
Site Plan	1 story square, 43'9" × 43'9"	78% of Texas SFD homes are 1 story per 2009 Residential Energy Consumption Survey (RECS) [15] ; a square home is agnostic to orientation.
Bedrooms	3	Majority of SFD homes (53%) have 3 bedrooms.
Bathrooms	2	A plurality of SFD homes (41%) have 2 bathrooms.
Foundation	Slab-on-grade, no insulation	Majority (76%) of SFD homes have a slab.
Ceiling Insulation	For Air Infiltration measure R-22. For Ceiling Insulation measure: Base R-2.5, Change R-30.	The average ceiling/wall insulation level for homes existing before 1998 is R-20.51/10.94, per utility baseline studies. It is assumed that all homes built from 1998 on had an average of R-30/13, per International Energy Conservation Code (IECC) 2009 code requirements. Per [15] , 78% of Texas SFD homes are pre-1998, and 22% were built on or after 1998. Taking the weighted average U values of insulation, the result is an overall average of U-0.0882/0.0454, or R-11.3/22.0.
Wall Insulation	R-11.3	See above.
Window Area	210 square feet	Per [15], the average Texas home has 14 windows, assuming an average size of 3' × 5' that makes for 210 square feet of windows.
Air Infiltration	For Ceiling Insulation measure: 12.2 ACH50 For Air Infiltration measure: Base 12.2 ACH 50, Change 7.43 ACH50	Based on LBNL's ResDB [16] , US average of 0.61 Normalized Leakage (NL) rate for SFDs; per ResDB [16] 0.5NL = 10 ACH50, so 0.61 NL = 12.2 ACH50.
Window U-Value	0.78	Combined the prevalence of single, double, and triple paned glass in Texas SFDs from [15] (58/41/1%) with the average U and solar heat gain coefficient (SHGC) for each pane level from LBNL's RESFEN database [17] , excluding windows with high solar gain coatings.

Input	Value	Source
Window SHGC	0.56	See above.
Thermostat Settings	Heating: 71.3°F during the day when someone is home, 67.7°F during the day when no one is home, 69.8°F at night; Cooling: 74.1°F during the day when someone is home, 76.6°F during the day when no one is home, 73.9°F at night.	Weighted average reported thermostat set points from [15] . Times associated with these set points are assumed to be the same as those specified by Energy Star program in US.
Duct Losses	18% total loss	From LBNL's ResDB [16] . National average total duct leakage is 18% of air flow.
Air Conditioning	11.3 SEER	Result of combining the average age of central cooling equipment from [15] with annual shipment-weighted SEER values from the US DOE.
Electric Heater	COP of 1	Fundamental property of electric resistance.

TABLE 4: Summer peak demand reduction for various efficiency measures from different approaches.

	Ceiling Insulation Austin (kW)	Air Infiltration Austin (kW)	Indoor Lighting Austin (kW)	Outdoor Lighting Austin (kW)
Probabilistic Approach (20 hours)	2.089	0.341	0.062	0
Top 2 Hours of All Peak Months (170 hours)	1.531	0.257	0.087	0
Heat Wave (9 hours)	2.036	0.344	0.056	0
Average Over Peak Period (510 hours)	1.511	0.241	0.069	0

The second and third alternatives are consistent with the definitions adopted by some state regulatory authorities in the US, as discussed earlier in this paper. The Public Utility Commission of Texas formerly required the Average over Peak Period method. Note that two of these three methods ignore the weather information in the TMY file.

The peak demand reduction from two weather-sensitive efficiency measures, ceiling insulation type and air infiltration, is presented in Table 4. The estimated average summer demand reduction in Austin for a prototype home using the probabilistic analysis is 2.09 kW for the ceiling insulation efficiency measure and 0.34 kW for the air infiltration efficiency measure.

The demand reduction impacts of two non-weather-sensitive measures, indoor and outdoor lighting in Austin, have also been estimated. For indoor lighting kW savings, we assumed that 30% of the original usage would be saved if energy-saving indoor lighting equipment was installed. Thus an average of 0.062 kW savings could be calculated based on the Energy Gauge home simulation model during 20 summer peak hours. For outdoor lighting, we considered a variety of outdoor lighting equipment and assumed that 5 kW savings when the outdoor light is on is a reasonable deduction. Since none of the summer 20 peak hours occurs at night, the demand reduction associated with the outdoor lighting efficiency measure is 0 kW.

For the two weather-sensitive measures, the definitions involving the highest number of hours yield the smallest estimated peak demand reduction. This is a reasonable result, since including further hours (without regard to the temperature associated with those hours) into a calculation of average impacts shall lower the average and bias the results downward. Our proposed probabilistic method provides estimates which are very similar to the Heat Wave method for the summer. The impact of the indoor lighting efficiency measure is greatest under the Top 2 Hours of All Peak Months Definition.

Winter peak demand reduction estimates for our proposed approach can be implemented using steps similar to those described above. However, RelativeMaxTemp needs to be replaced by RelativeMinTemp to represent the ratio between the actual temperature in a central location within the ERCOT market (Austin) for a given interval and the lowest temperature reading during the winter in the year. A Heat Wave calculation is not performed for the winter peak. The winter kW savings estimated under three definitions appear in Table 5.

The probabilistic approach produces far higher (and more-realistic) winter peak impact estimates for the weather-sensitive efficiency measures. The wide difference in estimates using different approaches can be traced to Texas' climate. Freezing temperatures set the winter peak and are a relatively rare event in this southern state. Deep freezes follow no predictable pattern. That is, one would not expect them to predictably occur during the same month-of-year and hour-of-day year after year. Consequently, the Top 2 Hours of All Peak Months performs poorly. Averaging over a prolonged winter peak period (in this case, from 6 am to 10 am and 6 pm to 10 pm) performs very poorly, as well, since many hours with mild temperatures and no need for space conditioning would be introduced into any peak period average.

All 20 of the winter peak hours happen after sunset and before sunrise. Consequently, the demand reduction in the winter for outdoor lighting is 5 kW under two of the three definitions. A lower peak demand reduction estimate is obtained when some daylight hours are included in the definition, as under the Average Over Peak Period calculation.

TABLE 5: Winter peak demand reduction from different approaches.

	Ceiling Insulation Austin (kW)	Air Infiltration Austin (kW)	Indoor Lighting Austin (kW)	Outdoor Lighting Austin (kW)
Probabilistic Approach (20 hours)	2.253	0.810	0.134	5
Top 2 Hours of All Peak Months (124 Hours)	0.601	0.197	0.183	5
Average over Peak Period (510 Hours)	0.933	0.239	0.105	3.583

For indoor lighting, similar results are obtained under any of the approaches considered. For the weather-sensitive measures, the probability-based method provides far more-plausible results for a measure's impacts on winter peak for Texas. Extreme temperatures indeed largely coincide with peaks in energy use, so the impacts of an efficiency measure during extreme weather (with adjustments for the time-of-day and month-of-year) should be used when estimating winter peak demand impacts. The use of simple temporal patterns which ignore temperatures or the averaging over large numbers of hours is inappropriate. It is suspected that in a colder climate where heat waves are a relatively rare event, the naïve application of patterns (without regard for temperature) or averaging to obtain summer peak impacts would similarly lead to implausible results.

11.5 CONCLUSIONS

Utility system planners and energy efficiency program administrators are interested in the impacts of energy efficiency programs at the time of peak demand on a utility system or energy market. Yet, it is not obvious which hour(s) correspond with peak hours when the output from a building energy use simulation model solved with TMY data is examined. Should the hour associated with the highest (or lowest) temperature be used? Should an average of the measure's impacts during the hours and months within which the utility's peak typically falls be used? Should impacts when design conditions are experienced be used? Should impacts during consecutive days of extreme weather be averaged? Would a lot of averaging dilute the impact of a weather-sensitive measure?

This paper proposes a formal probabilistic method to address this problem. We select the hours in a TMY weather data file most likely to coincide with a peak hour, based on the temperature, hour-of-day, and monthof-year data contained within the TMY data file and the relationships between these variables and actual load data for a utility system. Logistic regression is used to estimate the relationships based on actual historical data. The estimated relationships and TMY data are used to calculate the probability that an hour represented in the TMY data file would be included among a set of peak hours.

Our proposed approach represents a considerable improvement over existing practices which estimate impacts based solely on extreme temperatures in a TMY file, estimate impacts based upon design-day conditions, averages impacts over a large number of hours within a "peak period", or relies upon typical times of peak occurrence without consideration of the temperature in the TMY file during those hours. When applied to data for Texas, a probability-based approach provides more-realistic estimates of winter peak impacts, relative to two alternatives.

When estimating the impacts of an efficiency measure upon summer peak demand, our approach provides impacts similar to the Heat Wave approach being used in California.

REFERENCES

1. US Environmental Protection Agency (2006) National Action Plan for Energy Efficiency. US Environmental Protection Agency. http://www.epa.gov/cleanenergy/documents/suca/napee_report.pdf
2. Itron, Inc. (2013) DEER Database: 2011 Update Documentation, Appendices. http://www.deeresources.com/files/DEER2011/download/2011_DEER_Documentation_Appendices.pdf
3. Northeast Energy Efficiency Partnerships (2013) Technical Reference Manual, Version 3.0. http://www.neep.org/Assets/uploads/files/emv/emv-products/TRM_March2013Version.pdf
4. New York Public Service Commission (2010) New York Standard Approach for Estimating Energy Savings from Energy Efficiency Programs: Residential, Multi-Family, and Commercial/Industrial Measures. http://www3.dps.ny.gov/W/PSCWeb.nsf/96f0fec0b45a3c6485257688006a701a/766a83dce56eca35852576da006d79a7/$FILE/TechManualNYRevised10-15-10.pdf
5. Public Service Commission of Wisconsin (2010) Focus on Energy Evaluation, Business Programs: Deemed Savings, 2010. http://www.focusonenergy.com/sites/default/files/bpdeemedsavingsmanuav10_evaluationreport.pdf
6. State of Illinois (2012) Energy Efficiency Technical Reference Manual. http://ilsagfiles.org/SAG_files/Technical_Reference_Manual/Illinois_Statewide_TRM_Version_1.0.pdf
7. Xcel Energy (2012) 2012/2013 Demand Side Management Plan, Docket No. 11A-631EG. http://www.xcelenergy.com/staticfiles/xe/Marketing/Files/CO-DSM-2012-2013-Biennial-Plan-Rev.pdf
8. New Jersey's Clean Energy Program (2007) Protocols to Measure Resource Savings. http://www.njcleanenergy.com/files/file/Protocols_REVISED_VERSION_1.pdf
9. Efficiency Maine (2013) Residential Technical Reference Manual. http://www.efficiencymaine.com/docs/EMT-TRM_Residential_v2014-1.pdf
10. Zarnikau, J. and Thal, D. (2013) The Response of Large Industrial Energy Consumers to Four Coincident Peak (4CP) Transmission Charges in the Texas (ERCOT) Market. Utilities Policy, 26, 1-6. http://dx.doi.org/10.1016/j.jup.2013.04.004
11. Train, K. (2003) Discrete Choice with Simulation. Cambridge University Press, New York. http://dx.doi.org/10.1017/CBO9780511753930
12. SAS Institute Inc. (1990) SAS/STAT User's Guide, Vol. 1 & 2, Version 6. 4th Edition, SAS Institute Inc., Cary.
13. Everitt, B. and Hothorn, T. (undated) A Handbook of Statistical Analyses Using R. http://cran.r-project.org/web/packages/HSAUR/vignettes/Ch_logistic_regression_glm.pdf

14. Florida Solar Energy Center (undated) Energy Gauge. http://www.energygauge.com/
15. US Department of Energy (2009) Energy Information Administration, Residential Energy Consumption Survey (RECS). http://www.eia.gov/consumption/residential/
16. Lawrence Berkeley National Laboratory. Residential Diagnostics Database. http://resdb.lbl.gov/
17. Lawrence Berkeley National Laboratory. RESFEN. http://windows.lbl.gov/software/resfen/resfen.html

CHAPTER 12

A New Energy Management Technique for PV/Wind/Grid Renewable Energy System

ONUR OZDAL MENGI AND ISMAIL HAKKI ALTAS

12.1 INTRODUCTION

In today's world, the increasing need for energy and the factors, such as increasing energy costs, limited reserves, and environmental pollution, leads the renewable energy to be the most attractive energy source. Since these sources have unlimited supply and they do not cause environmental pollution, they are studied extensively lately and utilized more and more every day. Governments put in new legislations and feed-in-tariffs to encourage the investors to install new renewable energy utilization sites [1–3] and studies on this topic are supported by many foundations.

Renewable energy sources consist of solar energy, wind energy, geothermal energy, and wave energy which are considered to be endless since they exist naturally and they always renew themselves [4]. It is one of the

A New Energy Management Technique for PV/Wind/Grid Renewable Energy System. © *Mengi OO and Altas IH. Current Status and Prospects of Biodiesel Production from Microalgae.* International Journal of Photoenergy ***2015** (2015), http://dx.doi.org/10.1155/2015/356930. Licensed under Creative Commons Attribution 3.0 Unported License, http://creativecommons.org/licenses/by/3.0/.*

important topics that researchers and scientist work on to obtain energy from these sources and use this energy by transforming it into the form of electrical energy.

Solar and wind energies have a distinguished place among these energy types. There are wind and sun everywhere on earth; therefore, there is more intense study on these sources. The aim is not only to obtain the energy but also to turn the energy to proper values, manage the existent energy, and terminate the harmonics. While managing all these, lowering the cost of the system in every step is taken into consideration. Today, producing electrical energy from these renewable sources appears to be the main objective [5–7].

The combined operation of these systems is far more complex than operating them separately. In a system with only solar or wind energy, just one element is controlled. In a hybrid scheme, both sources are controlled individually and simultaneously depending upon the operating conditions and energy demand. During low sunlight conditions, photovoltaic (PV) solar panel cannot supply consistent power. Similarly, wind turbine will not work in conditions without wind. In this case, the required energy must have the structure to make up the lack of energy in conditions when this system does not work regularly or the composition produces less energy than the requirement. Power management assures that the system works efficiently while preventing the lack of energy in loads. Here it is aimed at obtaining clean and sustainable energy in stable frequency and definite voltage. While or after obtaining the energy, harmonics must be definitely controlled.

Power management is important to assure both economical and efficient work of the system in combined usage of renewable energy sources. Variable weather conditions, day-night conditions, and rapid change in voltages make this necessary. Power management can be achieved by using maximum power point tracking (MPPT) [8] devices in order to determine the most efficient operating point of a system in a particular weather condition and by switching the systems so that they become active to support each other dynamically. It is important to keep the backup batteries full in times when there is neither sun nor wind. Without backup batteries there will be no energy in the system. In this case, for instance, it is computerized control mechanism's duty to link the system to the grid, connect the generator or determine, and manage the related situations.

Nowadays renewable energy sources are structured in two ways as grid connected and standalone. Renewable energy sources as solar energy and wind energy can be used to feed loads far from the grid especially the home type ones. However, there are problems in these types of systems when there is no sun or wind. Users become fully powerless after the batteries are flat which are used as backup systems. An alternative situation to this is to connect the loads to the grid if they are close to it, in conditions that there is no sun or wind and the batteries are empty [9].

In literature review, it can be seen that there are many researches which include wind turbine and PV solar panels used together [10, 11] and the loads are fed with the energy obtained and power management conducted [12–14]. The main aim here is to gain maximum power according to environmental conditions and whether the obtained energy is to be fed by wind energy system (WES) or by PV solar panels systems according to changing load conditions. Similar to wind turbines and PV solar panels, there are several studies related to energy management and power flow in electrical power systems and other energy generation units [15–17].

Energy sources such as PV solar panels, wind turbines, fuel cell, and diesel generator can be used both as standalone or hybrid. There are many studies and utilizations such as wind/PV [10, 11], wind/fuel cell [18], PV/battery [19], wind/battery [20], PV/wind/fuel cell [21], PV/fuel cell [22], PV/wind/battery [23], and PV/grid [24]. The studies aim to increase power quality, ensure energy sustainability, and stabilize the amplitude and frequency of the voltage on the load side on a definite value. Besides, the energy management occupies an important part of the studies related to renewable energy utilization schemes [13]. Energy management in renewable energy systems deals with both source and user side control issues to keep the overall system running smoothly [25, 26].

In addition to this, MPPT is one of the important parts of the work, because IEMS calculates maximum power in WEC system. There are various methods which produce MPPT to obtain maximum power from RES. It is tried to run the system continuously at this point by defining the maximum power point with more efficient controls of power electronics converters. While there are studies for calculation of instantaneous generated power decided according to measurements of the environmental conditions, studies which focus on the efficient controls are conducted for similarly used engines to produce maximum power production. It is aimed that the wind turbines work with maximum efficiency [27–33]. In this study, there is an MPPT designed in a different way from these methods. Here, smart control software continuously and accurately calculates the maximum power that can be obtained from the wind turbine. MPPT is an important part of IEMS. Moreover, it is different than the other methods which include expensive control and measurement methods in that it is much cheaper and simpler.

In this study, a power management system will feed the loads from a hybrid power generation system consisting of PV solar panels, WES and grid. WES consists of a different and new MPPT method. The hybrid system is connected to a common DC bus, which is used as a power pool for sustainability. PV system is also connected to a backup battery unit to be charged for emergency usage when additional power is needed. In addition to the source side, the load side management is also very important for the renewable energy systems and also considered in this study.

Besides, energy management software can rapidly and continuously respond without being bounded to environmental conditions, which keep continuously some amount of power in reserve and during instantaneous load changes control the system efficiently. This study is different from the others in its being efficient management approach and having different, cheaper, and simpler peak power point tracking.

12.2 SYSTEM DESCRIPTION AND METHODOLOGY

The overall scheme of the proposed hybrid renewable power management system is given in Figure 1. The system consists of PV and wind power generating units and a utility grid as hybrid electrical sources. These three generating units are connected to a DC power pool over the required converters. A backup battery group is also connected to PV system in order to store the extra generated solar power when all generated power from the PV is not delivered to the load. AC load types are considered in the system and they are connected to AC power bus, which is fed from the common DC power pool. Data collected from source side and load side is transferred to a computer to be evaluated for decision making process of the power management system.

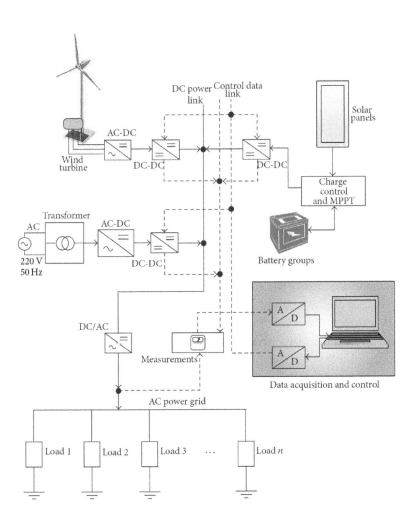

FIGURE 1: Renewable power generation system and energy management.

The power management algorithm (PMA) is developed to operate both photovoltaic energy system (PVES) and wind energy system (WES) at the maximum power they can generate under various environmental conditions while maintaining power supply demand of the load side at required amount. Therefore the power management includes maximum power point tracing of PVES and WES, energy storage, utility connection, and load switching. Besides, the power quality issues such as harmonic elimination, voltage sags, voltage increments, frequency deviations, and voltage magnitude are included in the management system. It is obvious that there are too many inputs and parameters into the management decision algorithms. A fuzzy logic based decision making algorithm is developed and used for this multi-input, multiobjective system.

12.2.1 PV SYSTEM

PVES is modeled using the well-known characteristic equation of PV cell. The voltage-current equation of a PV cell is based on photocurrent of a P-N junction semiconductor. The photocurrent is a function of solar irradiation and changes with the sunlight. The voltage across the connection terminal of the P-N device varies as a function of the photocurrent. When there is a path across the terminals of the P-N device, which is the photovoltaic cell, external current flows through this path, if there is a load on the path then through the load. The maximum value of this external current, or, in other words, load current, is the short circuit current and assumed to be equal to the generated photocurrent. It is observed that as the cell current increased, the cell gets heated resulting in a decreased terminal voltage. Therefore, considering this voltage decrement and reverse saturation current of the P-N diode, the terminal voltage of a single PV cell is written as in

$$V_C = \frac{A \times k \times T_a}{e} \ln\left(\frac{I_{PH} + I_S - I_C}{I_S}\right) - R_S \times I_C \qquad (2)$$

I_c: cell output current (A), I_{PH}: photocurrent, function of irradiation level and junction of temperature (A), I_s: reverse saturation of current of diode (A), V_c: cell output voltage (V), R_s: series resistance of cell, e: electron charge ($1.6021917 \times 10^{-19}$ C), k: Boltzmann constant (1.380622×10^{-23} J/$^{\circ}$K, T_a: reference cell ambient temperature ($^{\circ}$K), and A: curve fitting factor (100).

PV solar cells are connected in series and parallel combinations and manufactured as PV modules to be used more effectively in commercial applications. The voltage of a PV module is determined by the PV cells connected in series and the current of a PV module is determined by the number of parallel connected series branches. The required load power is then obtained by connecting the PV modules in series and in parallel yielding the PV arrays [24].

The power flow path from PV power generation unit to the load is shown in Figure 2. Four PV modules connected 2 in series and 2 series

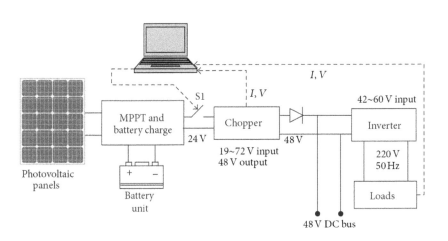

FIGURE 2: The PV power generation part of the system.

branches in parallel are used here to get an array with 42 volts and 5 amperes maximum under good weather conditions.

PV array and battery groups are connected to each other with a device including a battery charging regulator and maximum power point tracker. When the sunlight is not sufficient, the batteries step in and supply the necessary power to the loads. The MPPT is used to transfer the maximum generated power from the PV array and charge the batteries if any power is left after feeding the load. This unit is MOTECH PV4830 MPPT charge controller. The batteries are also charged from the DC power bus when the sunlight is insufficient. A battery charge regulator with 12/24/36/48 V and 30 A is used as a charging interface device. Total peak power generated by the PV array under good weather conditions is about 320 Wp. Since the value of the generated voltage from the PV array changes depending upon sunlight, a DC chopper (19~72 V DC input voltage) is used to keep the DC voltage from the PV panels at 48 V. This chopper is boost converter. The diode that is used after the chopper is located in order to protect the chopper. It is a kind of electronic fuse. It is a diode that prevents reverse current flow with a high current. This is the magnitude of the DC bus voltage, which is inverted to 220 V, 50 Hz AC voltage by a boost-up inverter. In this scheme, the current and voltage data from the loads are measured and transferred to the computer, besides the input voltage and output current of the chopper to be used in decision making process.

12.2.2 WIND TURBINE EMULATOR AND WES

The wind turbines convert the mechanical energy that is produced by the wind to electrical energy. To use this electrical energy a voltage and a frequency regulation has been needed. The model of the wind turbine is developed by the basis of the steady state power characteristics of the turbine.

Calculations of induction machines are completed in d-q axis frame. Figure 3 shows the depictions of axis frames.

Here we can obtain

$$\begin{bmatrix} V_{sd} \\ V_{sq} \end{bmatrix} = \begin{bmatrix} \cos\theta_s & -\sin\theta_s \\ \sin\theta_s & \cos\theta_s \end{bmatrix} \begin{bmatrix} V_{s\alpha} \\ V_{s\beta} \end{bmatrix} \tag{2}$$

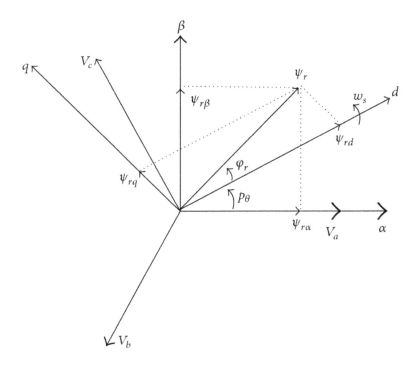

FIGURE 3: Phase depiction of components of axis frame in rotor flux vector.

After the necessary conversions are completed, d-q axis frame equivalent of the machine is obtained as shown in

$$
\begin{bmatrix} V_{sd} \\ V_{sq} \\ 0 \\ 0 \end{bmatrix} = \begin{bmatrix} R_S & 0 & 0 & 0 \\ 0 & R_S & 0 & 0 \\ 0 & 0 & R'_r & 0 \\ 0 & 0 & 0 & R'_r \end{bmatrix} \begin{bmatrix} i_{sd} \\ i_{sq} \\ i_{rd} \\ i_{rq} \end{bmatrix} + \begin{bmatrix} L_S & 0 & L_M & 0 \\ 0 & L_S & 0 & L_M \\ L_M & 0 & L'_r & 0 \\ 0 & L_M & 0 & L'_r \end{bmatrix} \frac{d}{dt} \begin{bmatrix} i_{sd} \\ i_{sq} \\ i_{rd} \\ i_{rq} \end{bmatrix}
$$

$$
+ \begin{bmatrix} 0 & -\omega_s L_S \omega_s L_m & 0 & \\ \omega_s L_S & 0 & 0 & \omega_s L_M \\ \omega_r L_m & 0 & 0 & -\omega_r L_r \\ 0 & \omega_r L_m & \omega_r L_r & 0 \end{bmatrix} \begin{bmatrix} i_{sd} \\ i_{sq} \\ i_{rd} \\ i_{rq} \end{bmatrix} \tag{3}
$$

$$M_e = pL_m\left(i_{sq}i_{rd} - i_{sd}i_{rq}\right) = J\frac{d^2\theta}{dt^2} + B\frac{d\theta}{dt}$$

(4)

Angular frequency ω_s is as shown below:

$$\omega_s = \omega_r + p\omega$$

(5)

Here ω_s represents the angular frequency of stator fluxes and ω_r represents rotor fluxes angular frequency and angular speed of machine shaft, also represented by

$$\omega_s = \frac{d\theta}{dt} = 2\pi\frac{n_s}{60}$$

(6)

and n_s is synchrony speed. ω is shown in

$$\omega = \frac{d\theta}{dt}$$

(7)

And n_s is as in

$$n_s = 60\frac{f_s}{p}$$

(8)

Here f_s represents the stator flux value.

    The relation between flux and current in d-q axis frame is as in (9) and (12):

$$\psi_{sd} = L_s i_{sd} + L_m i_{rd}$$

(9)

$$\psi_{sq} = L_s i_{sq} + L_m i_{rq}$$

(10)

$$\psi_{rd} = L'_d i_{rd} + L_m i_{sd} \tag{11}$$

$$\psi_{rq} = L'_r i_{rq} + L_m i_{sq} \tag{12}$$

The state-space model according to (0,d,q) stator and rotor axis frames of the system can be seen in (13) and (17) [34]:

$$\frac{di_{sd}}{dt} = \frac{1}{\sigma L_s}\left[-R_E i_{sd} + \sigma L_s \omega_s i_{sq} + \frac{L_m R'_r}{L'^2_r}\psi_{rd} + p\omega \frac{L_m}{L'_r}\psi_{rq} + V_{sd}\right] \tag{13}$$

$$\frac{di_{sq}}{dt} = \frac{1}{\sigma L_s}\left[-R_E i_{sq} + \sigma L_s \omega_s i_{sd} - p\omega \frac{L_m}{L'_r}\psi_{rd} + \frac{L_m R'_r}{L'^2_r}\psi_{rq} + V_{sq}\right] \tag{14}$$

$$\frac{d\psi_{rd}}{dt} = \frac{R'_r L_m}{L'_r}i_{sd} - \frac{R'_r}{L'_r}\psi_{rd} + \omega_r\psi_{rq} \tag{15}$$

$$M_e = p\frac{L_m}{L'_r}\left(i_{sq}\psi_{rd} - i_{sq}\psi_{rq}\right) = J\frac{d\omega}{dt} + B\omega \tag{16}$$

$$M_e = p\frac{L_m}{L'_r}\left(i_{sq}\psi_{rd} - i_{sq}\psi_{rq}\right) = J\frac{d\omega}{dt} + B\omega \tag{17}$$

Consider the following symbols:
- L_s: stator winding inductance (H),
- L_r: rotor winding inductance (H),
- M_m: maximum mutual inductance between rotor and stator (H),
- R_s: stator phase resistance (Ω),
- R_h: circle peace resistance between two strips (Ω),
- R_c: strip resistance (Ω),
- M_{ss}: converse inductance between stator phase windings (H),
- M_{rr}: mutual inductance between rotor strips (H),
- μ_o: $4\pi 10^{-7}$,
- g: air gap (m),

- A: air gap segment (m^2),
- p: number of pole pairs,
- ω_s: stator angular frequency,
- ω_r: rotor angular frequency,
- ω: synchronous speed,
- f_s: stator frequency,
- ψ_s: stator flux vector,
- ψ_r: rotor flux vector,
- θ: machine axis rotation angle,
- J: viscosity moment,
- B: viscosity friction coefficient.

The WES emulator is established by coupling two squirrel cage asynchronous machines together as seen in Figure 4. The power of the machine to be used as the prime mover for representing the wind turbine is selected as 5 kW and the power of the other machine used as the generator is selected as 3 kW so that the generator can also be operated at overpower conditions to expand the analysis of the operating cases.

The machine on the left is a prime mover representing the wind turbine and is controlled with a V/f speed controller. Depending on the wind speed, the output voltage of the generator, the second machine from the left, changes between 320 V and 400 V. Three phase voltage magnitudes obtained from the generator are reduced to a lower level using a step down transformer with a ratio of 380 V/36 V. Therefore, transformer output voltages change between 30 V and 38 V. The output voltage magnitude of 3-phase full bridge diode rectifier is calculated as in

$$U_O = 1.654 \cdot R_{R_{MAX}} \tag{18}$$

where U_o is rectifier output voltage (V) and U_{Rmax} is the maximum value of single phase voltage (V). Full bridge rectifier output voltage changes from 40,5 V to 51 V. A DC chopper is used to convert this variable voltage to a 48 V constant DC to be connected to common DC bus, which has a 48 V constant DC voltage.

The reactive power to generate energy at this point is supplied with a condenser group. Later on, this three-phase voltage which is rectified is sent through a chopper and it is brought to 48 V value and connected to common DC bus. Then an inverter is used to convert this 48 V voltage to 230 V/50 Hz alternating voltage and the loads are fed. Here the current and voltage information on the loads and copper input voltage and output current is measured and these values are transferred to the computer.

12.2.3 GRID CONNECTION

The proposed renewable energy scheme consists of two-type operating conditions. The first case described in previous sections is the one supplying power to individual loads where the utility grid is not available or not connected. The second case deals with utility connected renewable energy scheme as shown in Figure 5. Hence power generated from each power generation system is collected in 48 V direct current bus.

As shown in Figure 4, the utility voltage is converted to 48 V DC constant value to be connected to 48 V common DC bus as it is done for wind and PV systems.

Data collected from various parts of the system is transferred to the computer and used for control and decision making processes. The interfacing between the real system and the computer is established by NI USB 6259 data acquisition card.

12.2.4 POWER MANAGEMENT AND SUSTAINABILITY

The electrical power generated by renewable sources such as wind and solar power is affected by environmental conditions resulting in problems in load part. When there is no sun or the weather is cloudy, the power amount to be generated by solar energy changes. Accordingly, wind does not blow at the same speed all the time; it is discontinuous. Henceforth, energy amount to be generated from these sources are variable.

In particular in low powered applications, for instance, while supplying energy to a house from these sources, it is a problematic situation to have a power cutoff while watching a TV. Power sources must be efficiently operated in order to avoid such situations.

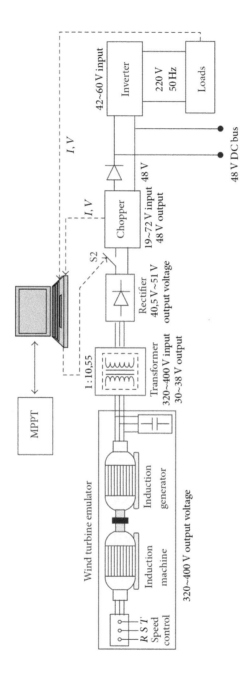

FIGURE 4: Feeding the loads with power generated from WES model.

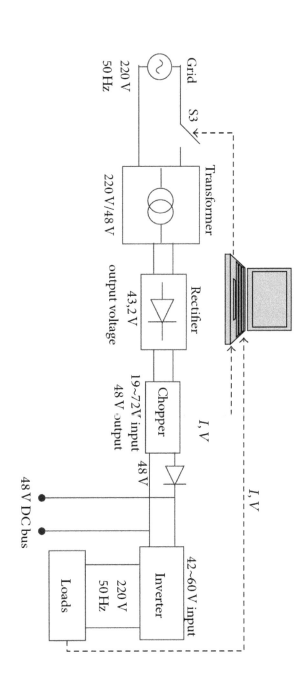

FIGURE 5: Grid connection structure in the system.

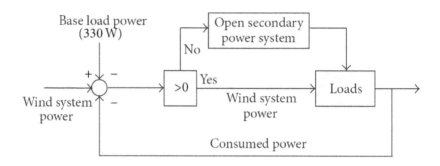

FIGURE 6: Basic block diagram of power management system.

Proposed PMA and decision making process are developed to prevent problems like voltage sags and discontinuities that occur due to either weather changes or sudden load changes. The intelligent decision making algorithm manages the energy storage and usage switching patterns so that energy sustainability is guaranteed.

General block diagram of the power management scheme is given in Figure 6. The total generated power from the wind is used as the primary supply power, which is used to supply base load power. As long as the wind power is sufficient, the secondary power system is kept off and additional generated power is stored. When the wind power is less than the required load power then the secondary power system, which is solar and/or utility grid, is activated. The service order of the secondary power system goes as solar PV, storage, and utility grid. Since the utility grid requires additional payment, it is put at the end of the list and the priority is given to the wind and PV arrays [35]. The load power is calculated as in

$$P_{load} = P_{windsystem} + 300W \tag{19}$$

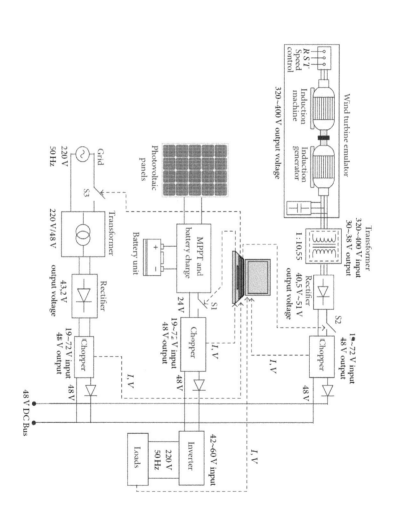

FIGURE 7: PV solar panels-WES-grid system.

12.2.5 POWER MANAGEMENT SYSTEM DESIGN

In order to solve sustainability and power quality problems, the power transfer from the renewable sources to load must be managed in a proper way. Therefore a PMA system has been designed to prevent power discontinuity and overvoltage and undervoltage operations so that the loads operate properly. The power management system is automated in an efficient way by switching on or off the sources and backup units. For example, if the wind power is sufficient enough to feed the load, then there is no need for the auxiliary sources of PV, backup batteries, and the utility. If the wind power decreases, the gap is filled by PV first, then batteries, and then the utility. The overgenerated power is stored and used only when needed.

The overall energy generation system established experimentally can be seen in Figure 7. In this system, the electrical power is generated by wind generator and PV solar panels. The utility is reserved as an auxiliary source to be used when needed. The power from the PV system is used to supply power to the load when the wind power is not sufficient and to charge the batteries when there is sufficient wind and sun power. Data collected from various parts of the overall system is transferred to the computer to be analyzed.

The main objective of employing a power management algorithm in power systems where the renewable energy is the priority supply is to have the power ready to be used and feed the load continuously. For this reason, the peak power value from both WES and PV solar panels must be calculated. MPPT device used for PV solar panels handles this duty on its own. Therefore, there is no need to calculate the peak power calculation in PV solar panels. MOTECH PV4830 MPPT charge controller calculates MPP by itself. However, the peak power value to be obtained from WES must be defined. The priority supply is WES in this system. At the same time, since the changing environmental conditions affect the amount of energy to be produced, power management is planned to avoid this effect by leaving a base power in the system. The base power is the power that must be supplied all the time for the loads with nonstop operating behaviors. The base power in the proposed system is defined as 300 W, which can be easily changed inside the software if desired. The system is designed to feed maximum 1 kW, which is more than three times the base power. If the environmental conditions are sufficient, which means there is enough sun and wind, wind energy generation system and PV solar panels generation system produce the maximum power they can, and the installed power can rise over 2 kW value. The main operational principle of the system is summarized as follows.

1. Initially the system is started with both solar and wind energy in service.
2. After the transients are over and measurements are done, WES or PV solar panels system will be kept working according to load condition. If the environmental conditions are not suitable for PV solar panels or WES to operate individually, both will operate. When the present condition does not meet the energy requirements of the load, grid will.
3. If WES can handle the load power requirement alone, it will operate. PV solar panels will be used to charge the batteries only.
4. If WES does not generate sufficient power, PV solar panels will engage and both will operate together. If both are insufficient, then the grid will engage.
5. If there is no wind, PV solar panels will feed the loads.
6. When there is no sun and the batteries are empty or the battery cannot feed the loads with its actual power amount, the grid will begin to operate. These steps bring up the importance of the following:
    (1) the operating time of each unit,
    (2) turnoff time of each unit,
    (3) the amount of load at present conditions.

Since only the wind energy will operate constantly, these lists of rules are processed by taking the measurements from wind energy system into consideration.

1. 300 W power will be supplied as the permanent power in the system. The wind turbine emulator is operated at various speeds in order to represent the generated wind power properly. The emulator has been set up so that the generated power will not be sufficient with the base power under a speed of 35 Hz. During this condition, solar energy will automatically begin to operate.

2. $V_{DC/DCG}$ chopper input voltage value is different in asynchronous motor driver frequencies. For instance, when the load power is 500 W, the generated voltage becomes 47.3 V at 44 Hz and 45.4 V at 43 Hz. It will decrease other value down to 31.5 V at 37 Hz. The relationship between voltage and frequency is used to estimate the operating frequency.

3. Considering both $I_{DC/DCO}$ and $V_{DC/DCG}$ values, asynchronous motor driver frequency, power dissipated at that moment, and maximum load values at that driver frequency can be detected. This case is also used to determine switch on or switch off times of PV solar panels. It should not be forgotten that 300 W power will always be in the system as the base power to be supplied.

4. Similarly when the PV solar panels are on, the generated voltage from the PV panels to the chopper decreases as load power increases. Using the measured values from this operating case, it is possible to find out how much power should be generated by the panels to feed the load. Chopper input voltage value will decrease as the batteries discharge. The batteries are assumed to be insufficient to feed the load when their voltage drops down below 21 V. If the sunlight is not enough to generate the required power to the load, the batteries also will not be usable with an output voltage less than 21 V.

5. While PV solar panels are feeding the loads, after the WES begins to operate, some of the required power is supplied from WES. If WES can supply all of the load power, then PV solar panels are switched to charge the batteries.

6. It is important to get measurements of these operating cases continuously in order to respond immediately in sudden load changes. The base power demand is 300 W. If a sudden load change increases over 300 W, the system is disconnected. For example, when the load power is 500 W when WES can give maximum 500 W, PV solar panels must operate so that in a 300 W sudden load increase the loads can be fed without the system failure. This action is taken to prevent system shutdown in power changes resulting from sudden charges according to the environmental conditions at the time.

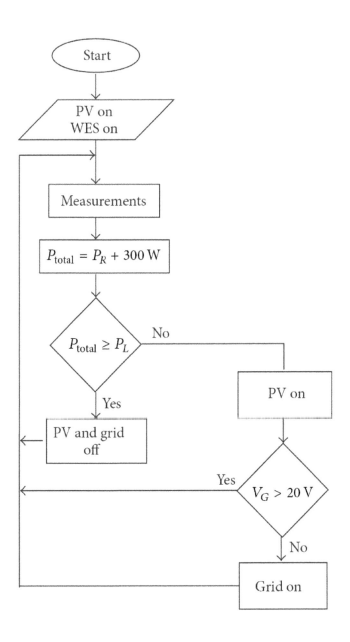

FIGURE 8: Simplified flow diagram of power management program.

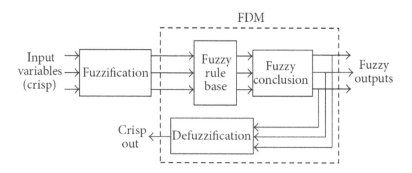

FIGURE 9: Fuzzy decision maker.

7. The grid connection is established when neither the sun nor the wind are sufficient enough to supply 300 W base power demand.

Flow diagram of power management program is given in Figure 8.

12.2.6 FUZZY LOGIC REASONING (FLR) AND CALCULATION OF PEAK WIND POWER

Fuzzy logic has so many applications in industry. Fuzzy logic reasoning (FLR) algorithms are generally used in fuzzy decision makers (FDM), which find applications in systems that require a conclusion from uncertain input data. Since the wind conditions are not certain and are not easily predictable, the power generated by the wind energy system becomes uncertain including the maximum generated power as well. Therefore a fuzzy reasoning algorithm is developed to determine the maximum power generated by WES. The fuzzy reasoning applied to determine the maximum power of the WES is represented in Figure 9. Fuzzy decision maker's MATLAB/Simulink model is given in Figure 10.

A FDM usually gets fuzzy inputs and evaluates them in the rule base system, which is set up earlier representing the input-output relations of the uncertain system in terms of fuzzy membership functions and fuzzy rules (FR). The FR is the evaluation of the rules to yield fuzzy conclusions from fuzzy inputs-fuzzy rules interactions, if the input variables are crisp. Certain input values are rated here. Thus, these values can be included in a value range that can be used by the controller and then it can be expressed verbally. In addition, it becomes a member of a group that has clear boundaries. Then they are fuzzified to fuzzy values. Similarly if the output is required as crisp value, then the concluded fuzzy outputs are converted to crisp values by the process called defuzzification.

In this study, the input values to the FDM are the current and voltage measured from the WES. The generated rules are used to relate the input voltage and current with the power of the WES depending upon the wind speed conditions. FR algorithm in the FDM is used to obtain maximum power generation from WES for uncertain and unpredictable speed conditions. The designed FR based FDM is modeled in Matlab/Simulink environment as shown in Figure 10. The maximum wind power value is determined by FDM using chopper's input voltage $V_{DC/DCI}$ and the output current is $I_{DC/DCO}$. The wind turbine, a 3-phase transformer, a 3-phase bridge rectifier, and a DC/DC converter are all assumed as a whole system. All of the calculations are based on the values of $V_{DC/DCI}$ and $I_{DC/DCO}$. Since the chopper output voltages are kept constant at 48 V DC, the main variable at the output terminals of the choppers or at common DC bus is the current at the output terminals of the choppers. The output current of the chopper used to control the voltage of WES is the variable that reflects the changes on WES power. Therefore the active power generated by the WES is obtained from

$$P_{DC/DC} = 48V \times I_{DC/DCO} \tag{20}$$

where 48 V is the chopper output voltage.

All of the MPPT calculations are based on the values of $V_{DC/DCI}$ and $I_{DC/DCO}$. As the wind speed changes, produced power values change. Therefore, current and voltage values that belong to system have changed too. These changes are followed with the chopper and its values $V_{DC/DC1}$ and $I_{DC/DC0}$. By following these two values the peak power value of the system's present speed value has been determined in this way. Those loading experiments were made for all wind speed values first, and then power for each speed was determined and transferred to the controller by blurring them. Controller uses abovementioned experience and control the system properly. It is a kind of expert system behavior.

The online data collected and transferred to computer is used to determine the amount of load power demand that supplied from the PV/wind sources. In the meantime the data representing the WES quantities are used by FDM to determine the maximum power generated by the WES for the instant the measurements made. The maximum power values of both WES and PV panels are used in power management part of the study. As seen in Figure 11, the crisp current and voltage values are converted to fuzzy values. Triangle type membership functions are used to represent 7 fuzzy subsets. In the system the current values and voltage values are varying from 3,5 A to 23,5 A (from I_1 to I_6) and from 27 V to 55 V (from V_1 to V_6), respectively. The lower limit of the voltage corresponds to the lower limit of the wind speed. Below the lower limit of the voltage the required energy conversion is not satisfied. Therefore the voltages below 27 V are treated as zero. The output of the FDM is the output power space (from W_1 to W_6) that calculates the maximum power value provided by the WES. The power generated by the WES varies as a function of the wind speed. Therefore the maximum power was obtained from the WES changes depending on the wind speed levels and must be determined for different speed levels as the wind speed changes. Therefore the maximum power surface of the WES is portioned into seven fuzzy subsurfaces as shown in Figure 12 [24].

Experimental test system is seen in Figures 13 and 14.

In WES, MPTT is implemented as software. There is not any e lectronic intervention to the generator. In addition to this, there is not any added hardware. For this reason, it is different from other MPTT systems.

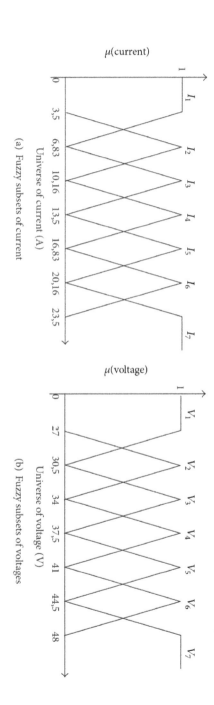

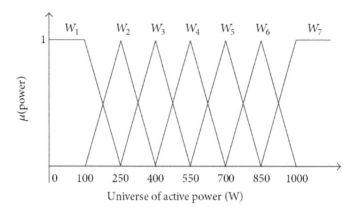

FIGURE 12: Fuzzy subset of power output spaces.

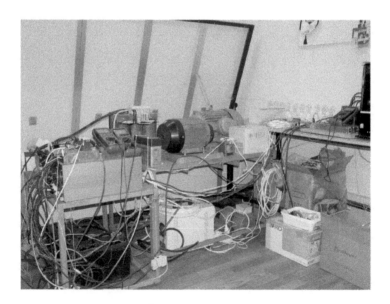

FIGURE 13: Appearance of experimental system on one side.

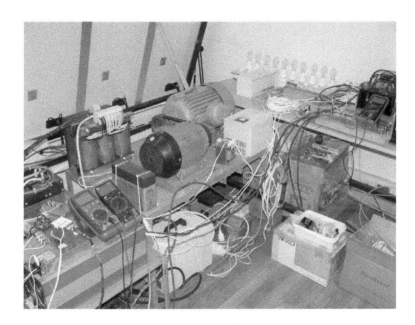

FIGURE 14: Appearance of experimental system from above.

12.3 TEST AND RESULTS

In Figure 15, the system consisting of WES and PV solar panels and without energy management software can be seen. The system fully operates in free mode. There is neither supervision nor a control mechanism.

As the first operating case, the system is analyzed when both wind and PV solar panels are on without applying any power management. For this case, chopper input voltage, V_R, in WES, chopper output current, I_R, PV solar panel voltage, V_G, to the chopper input, chopper output current, I_G, load voltage, V_Y, and load current, I_Y, can be seen in Figures 16–18.

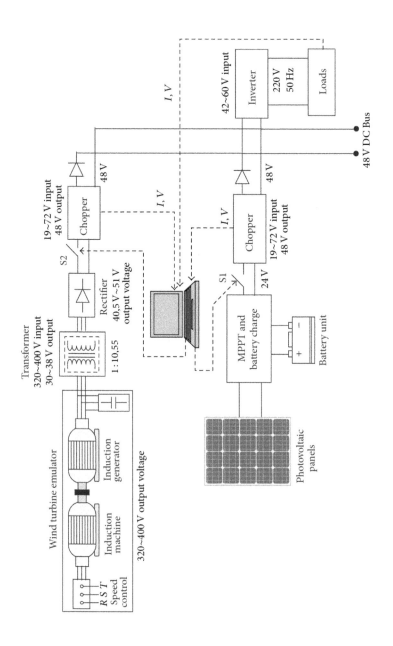

FIGURE 15: Uncontrolled wind-solar energy generation system.

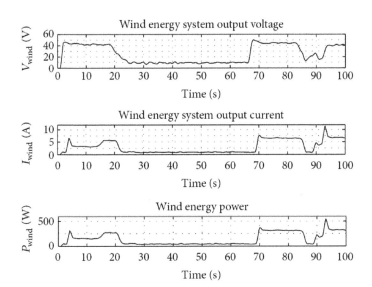

FIGURE 16: Voltage, current, and power variations of WES for case 1.

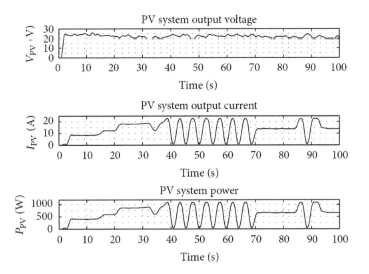

FIGURE 17: Voltage, current, and power change of PV solar panels for case 1.

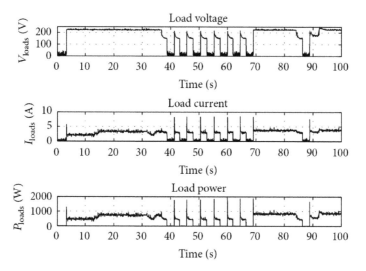

FIGURE 18: Voltage, current, and power change of loads for case 1

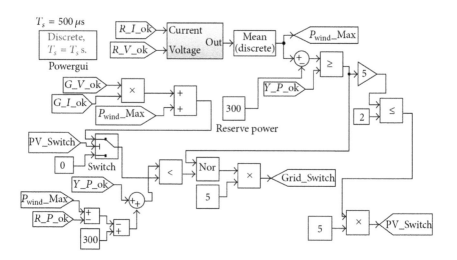

FIGURE 19: Simulink model of the proposed power management system.

It can be seen that when a load power of 800 W is on, WES and PV system together cannot feed the loads from 40 s to 70 s and from 86 s to 89 s, since they could not operate properly. Besides, the load voltage decreases down to 0 sometimes instead of remaining at the required value of 220 V. Although there is enough power generated in the system, power discontinuities occur. Actually the power generated PV solar panels can be used to eliminate the power discontinuities due to changes in wind speed by applying proper power management algorithms. In this case, a decision making system is needed to control the system and make decisions in order to avoid lack of power discontinuity on loads.

The proposed power management algorithm is realized in MATLAB/ Simulink environment using dynamic operational blocks library as shown in Figure 19.

The Simulink diagram in Figure 18 consists of the following subsystems:

- R_I_ok: wind system current,
- R_V_ok: wind system voltage,
- R_P_ok: wind system power,
- G_V_ok: PV system voltage,
- G_I_ok: PV system current,
- Y_P_ok: loads power,
- Pwind_Max: calculated maximum wind power,
- Grid_Switch: grid system switch (on/off),
- PV_Switch: PV system switch (on/off).

Figure 20 shows current, voltage, and power changes of WES for operating case 2 in which, the load power is changed arbitrarily in different time periods resulting in changes in the quantities of WES. It can be observed that WES is always active, yet, because of wind speed and powers extracted, the power amount transferred to the system changes.

Current, voltage, and power changes for PV system during case 2 can be seen in Figure 21. PV power is supplied to the load between the durationst = 20–75 s and t = 88–100 s.

The changes of current, voltage, and power from the utility grid are given in Figure 22. The grid transfers energy to the loads during the time interval of t = 45–68 s.

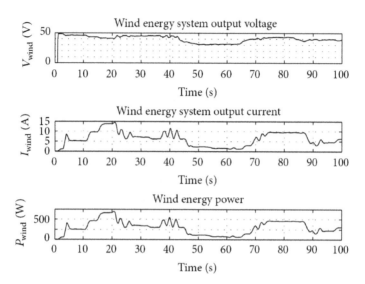

FIGURE 20: Current, voltage, and power change in WES for case 2.

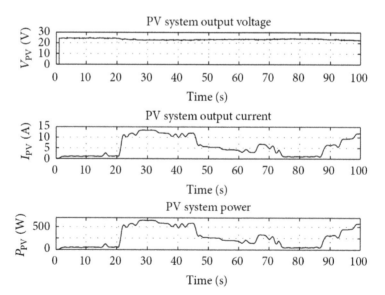

FIGURE 21: Current, voltage, and power change of PV system in case 2.

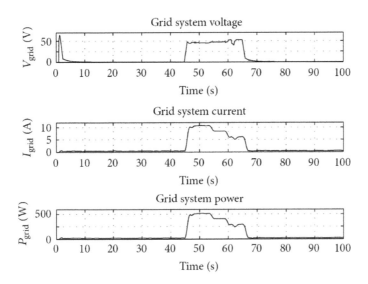

FIGURE 22: Variations of current, voltage, and power from the utility grid in case 2.

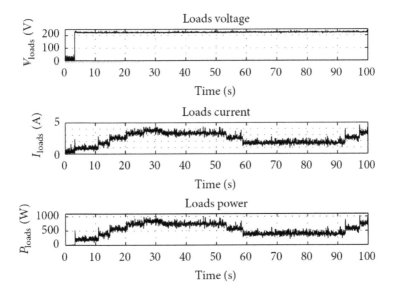

FIGURE 23: Current, voltage, and power change on the load in case 2.

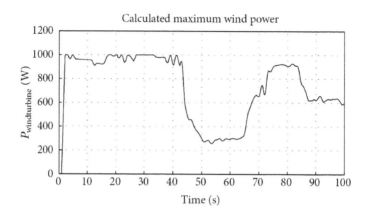

FIGURE 24: Peak power changes obtained from WES using FLR algorithm.

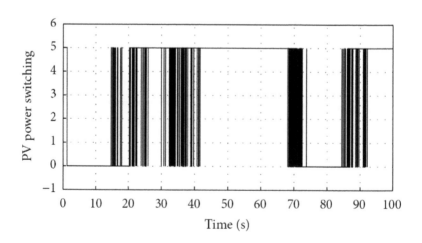

FIGURE 25: Switching instants of PV system when wind energy is not sufficient.

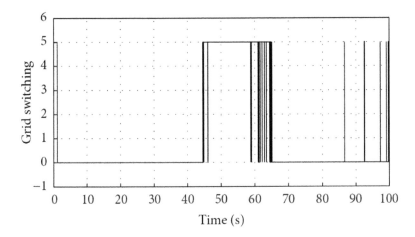

FIGURE 26: Switching instants of utility grid when wind and solar energy are not sufficient.

Current, voltage and power changes on the load terminals can be seen in Figure 23. The load amount changes constantly. There is a power consumption varying between 0 W and 900 W. 100 W lamps are used as load. Voltage on the load is 220 V. Varying amounts of current are taken from the energy generation system according to the load condition.

In Figure 24, variation of the power generated by WES is given. A fuzzy logic reasoning (FLR) algorithm is employed to manage this power.

In Figures 25 and 26, it can be observed that PV and grid system operates and switch on and off. During the time interval of t = 41–68 s the PV system is on while during the time interval of t = 45–58 s the utility grid system feeds the loads. Both systems operate for short time periods from time to time depending upon the sudden changes.

In Figure 27, we can see the details of voltage change on the loads. Wave shape is very close to sine. Distortions are observed since data is generally taken from voltage detector.

Between Figures 28 and 37, an asynchronous motor is added to lamps which are used as load and the system is restarted and then the results are drawn. Thus, by adding resistive load and also inductive load to the system, the behavior of the system under different load conditions is examined in detail.

The current, voltage, and power value changes from WES can be seen in Figure 28. The magnitude of the generated voltage is very low in time interval from 50 s to 70 s due to high reduction in wind speed.

Time variations of voltage, current, and power of PV system are given in Figure 29. PV solar panels operate especially in period from 45 s to 70 s. PV solar panels are active during this period because WES is not sufficient enough to generate the required load power.

Figure 30 shows the variations of voltage, current, and power of the utility grid. During the period from 45 s to 75 s the utility grid supplies power to the loads since both wind and PV systems are off or they do not generate required power due to low speed and/or low sunlight.

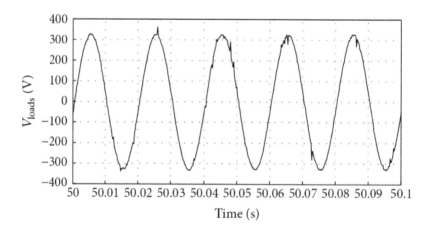

FIGURE 27: Wave shape of the voltage on loads.

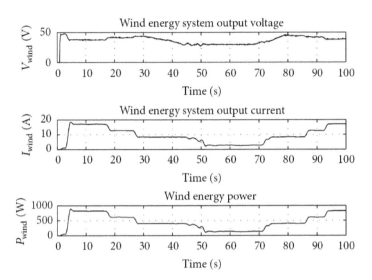

FIGURE 28: Current, voltage, and power changes of wind energy system.

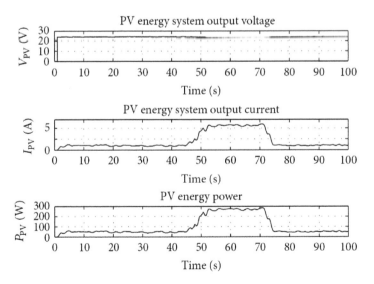

FIGURE 29: Current, voltage, and power change of PV energy system.

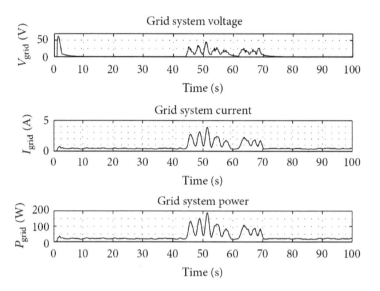

FIGURE 30: Current, voltage, and power change of grid system.

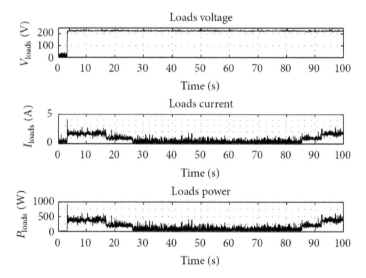

FIGURE 31: Current, voltage, and power change on the loads.

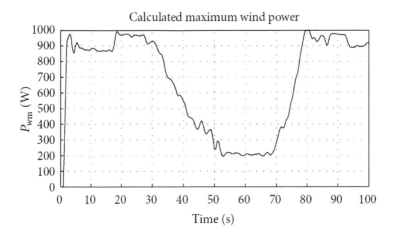

FIGURE 32: The maximum power change tracking of the WES by FLR.

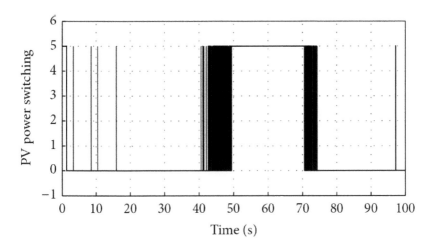

FIGURE 33: On and off switching pulses of PV panels.

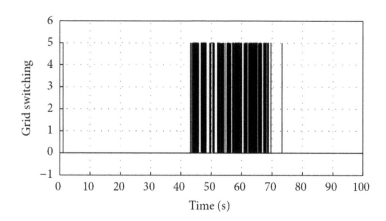

FIGURE 34: On and off switching pulses of utility grid.

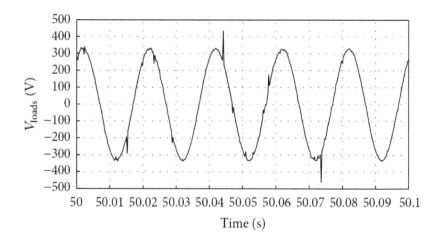

FIGURE 35: Voltage wave shape on the load.

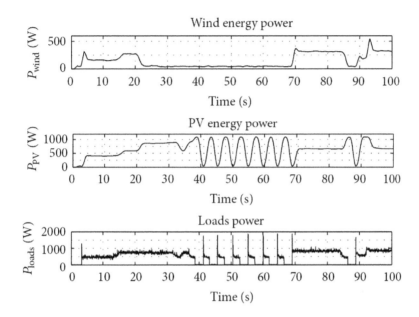

FIGURE 36: Power changes in the system without power management.

Current, voltage, and power changes on the load can be seen in Figure 31. Asynchronous motor load is always kept operating. During the time interval of t = 3–18 s, 200 W lamps are added to the system as loads in addition to the asynchronous motor. Later, in period t = 27–85 s, the lamps are turned off leaving only the asynchronous motors as the load. Meanwhile, wind speed is changed constantly. Load voltage is kept at 220 V, while the load current varies according to load condition.

The maximum power drawn from the WES is given in Figure 32. The generated power by WES is high around 900–950 W when the wind speed is high and it is low around 200 W when the wind speed is lower during the period from 50 s to 70 s.

The on and off switching instances of PV solar panels and utility grid system are shown Figures 33 and 34, respectively. In period of t = 40–70 s,

PV panels are on and, in period of t = 43–70 s, the grid system is on. They are turned on in order to support WES so that the loads will not be lacking of energy in that time period.

Wave shape details of voltage on the load are given in Figure 35. Although elements like asynchronous motor lead to harmonics, a very clear sine wave with a little amount of distortion is obtained.

The variations of power from wind, PV, and load busses are given in Figure 36 for case 1 where there is neither control nor power management algorithm. Since there is no power management, the load tries to get the power from PV and backup system, which are not sufficient enough to feed the load. Therefore load does not get the required power to operate.

Variations of the power from WES, PV, and utility grid along with load power and available maximum value of the wind power are shown in Figure 37. Proposed power management algorithm (PMA) has been applied for this case. Therefore there is no problem on supplied load power. As load or environmental conditions change, the load power is supplied from the sources in the order of priority use as wind, PV, and utility grid. If the power from the wind is sufficient for the load, the generated PV power is stored in backup batteries. If the wind power is not sufficient, then PV power is connected to complete the required power. If both wind and PV do not generate the required load power, then the utility grid is connected.

12.4 CONCLUSIONS

An intelligent energy management system (IEMS) for maintaining the energy sustainability in renewable energy systems is presented in this study. A renewable energy system consisting of wind and PV panels is established and used to test the proposed IEMS. Since the wind and PV sources are not reliable in terms of sustainability and power quality, a management system is required for sustainability on the load side. The proposed PMA is used to collect power from renewable sources and utility grid at a common DC bus and feed the loads preventing any power discontinuity. By employing PMA, a base power is always supplied to DC bus to be used by the loads that are operating permanently. Besides, PMA handles the effects of the changes in wind speed, solar irradiation, and amount of the load by operating wind, PV, and utility grid accordingly using intelligent decision making abilities. The proposed intelligent PMA is also used to determine and track the generated and available maximum wind power from WES so that the efficiency of the installed units is increased. Using the generated and required power information from the wind/PV and load sides, the fuzzy reasoning based PMA generates the required operating sequences to manage the overall system power with the minimum requirement from the utility. The IPMS is also designed to operate the renewable energy systems as a part of power utility. Therefore the IEMS can also be considered as a smart grid operator in the proposed RES application. Proposed IPMS can be extended to be used in distributed power systems for providing decisions on power management during critical peak power instances and fast power demand changes.

SYMBOLS

- I_C: Cell output current
- I_{pH}: Photocurrent function of irradiation level and junction of temperature
- I_S: Reverse saturation of current of diode
- V_C: Cell output voltage
- R_S: Series resistance of cell
- e: Electron charge
- k: Boltzmann constant
- T_{pil}: Reference cell operating temperature
- A: Curve fitting factor
- $V_{DC/DCG}$: Chopper input voltage
- $I_{DC/DCO}$: Chopper output current
- V_R: WES chopper input voltage
- I_R: WES chopper output current
- V_G: PV solar panels system chopper input voltage
- I_G: PV solar panels system chopper output current
- V_Y: Load voltage
- I_Y: Load current
- R_I_ok: Wind system current
- R_V_ok: Wind system voltage
- R_P_ok: Wind system power
- G_V_ok: PV system voltage
- G_I_ok: PV system current
- Y_P_ok: Loads power
- Pwind_Max: Calculated maximum wind power

- Grid_Switch: Grid system switch (on/off)
- PV_Switch: PV system switch (on/off)
- P_{Load}: Load power
- $P_{WindSystem}$: WEC system power
- L_s: Stator winding inductance (H)
- L_r: Rotor winding inductance (H)
- M_m: Maximum mutual inductance between rotor and stator (H)
- R_s: Stator phase resistance (Ω)
- R_h: Circle peace resistance between two strips (Ω)
- R_c: Strip resistance (Ω)
- M_{ss}: Converse inductance between stator phase windings (H)
- M_{rr}: Mutual inductance between rotor strips (H)
- μ_o: $4\pi10^{-7}$
- g: Air gap (m)
- A: Air gap segment (m^2)
- p: Number of pole pairs
- ω_s: Stator angular frequency
- ω_ρ: Rotor angular frequency
- ω: Synchronous speed
- f_s: Stator frequency
- ψ_s: Stator flux vector
- ψ_r: Rotor flux vector
- θ: Machine axis rotation angle
- J: Viscosity moment
- B: Viscosity friction coefficient.

REFERENCES

8. P. Denholm, R. Margolis, T. Mai et al., "Bright future: solar power as a major contributor to the U.S. grid," IEEE Power and Energy Magazine, vol. 11, no. 2, pp. 22–32, 2013.

9. J. Von Appen, M. Braun, T. Stetz, K. Diwold, and D. Geibel, "Time in the sun: the challenge of high PV penetration in the German electric grid," IEEE Power and Energy Magazine, vol. 11, no. 2, pp. 55–64, 2013.

10. K. Ogimoto, I. Kaizuka, Y. Ueda, and T. Oozeki, "A good fit: Japan's solar power program and prospects for the new power system," IEEE Power and Energy Magazine, vol. 11, no. 2, pp. 65–74, 2013.

11. V. Quasching, Understanding Renewable Energy Systems, Earthscan, London, UK, 2005.

12. T. Ackermann, Wind Power in Power Systems, John Wiley & Sons, Chichester, UK, 2005.

13. I. Akova, Yenilenebilir Enerji Kaynakları, Nobel Yayin Dagitim, Ankara, Turkey, 2008.

14. C. Kocatepe, M. Uzunoğlu, R. Yumurtacı, A. Karakaş, and O. Arikan, Elektrik Tesislerinde Harmonikler, Birsen Yayınevi, Istanbul, Turkey, 2003.

15. T. Esram and P. L. Chapman, "Comparison of photovoltaic array maximum power point tracking techniques," IEEE Transactions on Energy Conversion, vol. 22, no. 2, pp. 439–449, 2007.

16. G. M. Masters, Renewable and Efficient Electric Power Systems, John Wiley & Sons, New York, NY, USA, 2004.

17. J. L. Bernal-Agustín, R. Dufo-López, J. A. Domínguez-Navarro, and J. M. Yusta-Loyo, "Optimal design of a PV-wind system for water pumping," in Proceedings of the International Conference on Renewable Energies and Power Quality, pp. 1–6, Santander, Spain, March 2008.

18. Y. Thiaux, J. Seigneurbieux, B. Multon, H. B. Ahmed, and D. Miller, "Single phase AC power load profile emulator," in Proceedings of the International Conference on Renewable Energies and Power Quality, Santander, March 2008.

19. V. Courtecuisse, J. Sprooten, B. Robyns, M. Petit, B. Francois, and J. Deuse, "A methodology to design a fuzzy logic based supervision of hybrid renewable energy systems," Mathematics and Computers in Simulation, vol. 81, no. 2, pp. 208–224, 2010.

20. Y. Chen and J. Wu, "Agent-based energy management and control of a grid-connected wind/solar hybrid power system," in Proceedings of the 11th International Conference on Electrical Machines and Systems (ICEMS '08), pp. 2362–2365, IEEE, Wuhan, China, October 2008.

21. D. Das, R. Esmaili, L. Xu, and D. Nichols, "An optimal design of a grid connected hybrid wind/photovoltaic/fuel cell system for distributed energy production," in Proceedings of the 31st Annual Conference of IEEE Industrial Electronics Society (IECON '05), pp. 2499–2504, November 2005.

22. X. Zhu, D. Xu, P. Wu, G. Shen, and P. Chen, "Energy management design for a 5kW fuel cell distributed power system," in Proceedings of the 23rd Annual IEEE Applied Power Electronics Conference and Exposition, pp. 291–297, Austin, Tex, USA, February 2008.

23. K.-S. Jeong, W.-Y. Lee, and C.-S. Kim, "Energy management strategies of a fuel cell/battery hybrid system using fuzzy logics," Journal of Power Sources, vol. 145, no. 2, pp. 319–326, 2005.

24. Z. Jiang, "Power management of hybrid photovoltaic—fuel cell power systems," in Proceedings of the IEEE Power Engineering Society General Meeting, pp. 1–6, June 2006.

25. M. Nayeripour, M. Hoseintabar, T. Niknam, and J. Adabi, "Power management, dynamic modeling and control of wind/FC/battery-bank based hybrid power generation system for stand-alone application," European Transactions on Electrical Power, vol. 22, no. 3, pp. 271–293, 2012.

26. S. Harrington and J. Dunlop, "Battery charge controller characteristics in photovoltaic systems," in Proceedings of the 7th Annual Battery Conference on Applications and Advances, pp. 15–21, Pasadena, Calif, USA, January 1992.

27. S. Eren, J. C. Y. Hui, D. To, and D. Yazdani, "A high performance wind-electric battery charging system," in Proceedings of the Canadian Conference on Electrical and Computer Engineering (CCECE '06), pp. 2275–2277, Ottawa, Canada, May 2006.

28. D. B. Nelson, M. H. Nehrir, and C. Wang, "Unit sizing of stand-alone hybrid Wind/PV/fuel cell power generation systems," in Proceedings of the 2005 IEEE Power Engineering Society General Meeting, pp. 2116–2122, San Francisco, Calif, USA, June 2005.

29. R. Contino, F. Iannone, S. Leva, and D. Zaninelli, "Hybrid photovoltaic-fuel cell system controller sizing and dynamic performance evaluation," in Proceedings of the IEEE Power Engineering Society General Meeting, pp. 1–6, Montreal, Canada, October 2006.

30. Q. Mei, W.-Y. Wu, and Z.-L. Xu, "A multi-directional power converter for a hybrid renewable energy distributed generation system with battery storage," in Proceedings of the CES/IEEE 5th International Power Electronics and Motion Control Conference (IPEMC 2006), pp. 1932–1936, Shanghai, China, August 2006.

31. I. H. Altas and O. O. Mengi, "A fuzzy logic controller for a hybrid PV/FC green power system," International Journal of Reasoning-Based Intelligent Systems, vol. 2, no. 3, pp. 176–183, 2010.

32. D. Petkovića, S. Shamshirbandb, N. B. Anuar et al., "An appraisal of wind speed distribution prediction by soft computing methodologies: a comparative study," Energy Conversion and Management, vol. 84, pp. 133–139, 2014.

33. O. O. Mengi and I. H. Altas, "A fuzzy decision making energy management system for a PV/Wind renewable energy system," in Proceedings of the International Symposium on Innovations in Intelligent Systems and Applications (INISTA '11), pp. 436–440, IEEE, Istanbul, Turky, June 2011.

34. I. Munteanu, A. I. Bratcu, and E. Ceangă, "Wind turbulence used as searching signal for MPPT in variable-speed wind energy conversion systems," Renewable Energy, vol. 34, no. 1, pp. 322–327, 2009.

35. V. Galdi, A. Piccolo, and P. Siano, "Exploiting maximum energy from variable speed wind power generation systems by using an adaptive Takagi-Sugeno-Kang fuzzy model," Energy Conversion and Management, vol. 50, no. 2, pp. 413–421, 2009.

36. T. Senjyu, Y. Ochi, Y. Kikunaga et al., "Sensor-less maximum power point tracking control for wind generation system with squirrel cage induction generator," Renewable Energy, vol. 34, no. 4, pp. 994–999, 2009.

37. M. Arifujjaman, M. T. Iqbal, and J. E. Quaicoe, "Maximum power extraction from a small wind turbine emulator using a DC-DC converter controlled by a microcontroller," in Proceedings of the 4th International Conference on Electrical and Computer Engineering (ICECE '06), pp. 213–216, Dhaka, Bangladesh, December 2006.

38. V. Calderaro, V. Galdi, A. Piccolo, and P. Siano, "A fuzzy controller for maximum energy extraction from variable speed wind power generation systems," Electric Power Systems Research, vol. 78, no. 6, pp. 1109–1118, 2008.

39. V. Agarwal, R. K. Aggarwal, P. Patidar, and C. Patki, "A novel scheme for rapid tracking of maximum power point in wind energy generation systems," IEEE Transactions on Energy Conversion, vol. 25, no. 1, pp. 228–236, 2010.

40. V. Agarwal, R. K. Aggarwal, P. Patidar, and C. Patki, "A novel scheme for rapid tracking of maximum power point in wind energy generation systems," IEEE Transactions on Energy Conversion, vol. 25, no. 1, pp. 228–236, 2010.

41. M. K. Sarıoğlu, M. Gökaşan, and S. Boğosyan, Asenkron Makinalar ve Kontrolü, Birsen Yayınevi, Istanbul, Turkey, 2003.

42. O. O. Mengi, Yenilenebilir Enerji Sistemlerinde Süreklilik için Akıllı Bir Enerji Yönetim Sistemi [Ph.D. thesis], Karadeniz Technical University, Trabzon, Turkey, 2011.

Figures 10 and 37 are not available in this version of the article. To view this additional information, please use the citation on the first page of this chapter.

CHAPTER 13

Implications of Diurnal and Seasonal Variations in Renewable Energy Generation for Large Scale Energy Storage

F. M. MULDER

13.1 INTRODUCTION

Dominant resources for renewable electricity generation are solar and wind power. Solar power is generally seen as having the largest global technical potential [1,2] while the latter is on an implementation track leading to a significant percentage of the global electricity production. In 2012, close to 280 GW installed wind power is reported worldwide and forecasts indicate up to 100% growth between 2011 and 2016. [3] In many countries GW installed power wind parks are being built or planned for installation in the next decade amounting to several tens of percents of the average electricity generation. By 2050 as much as half of the current total energy use (~525 EJ/year) may be generated by solar and windpower alone (~270 EJ/year). [1,2] These solar and wind power generating capabilities are grid connected in order to transport energy from (on/offshore) harvesting to utilization site. Depending on operation level, these grids can operate locally in the distribution grid or on larger scale in the high voltage transmission grid. In addition, long distance grid connections are becoming available and are being considered to extend over several thousands of kilometers, e.g., from north west Europe to northern Africa, or across USA and Canada. The rationale behind such large sized grids is to enable trade and transport of power, and to secure energy supply by averaging the local fluctuations in the instantaneous generated solar and wind power over extended areas. But what systematic variation in time of the output power of extended grids can be expected? And how can one cope with the variation, using energy storage on what scales? Here a model is presented to enable to make such estimates.

Depending on technological development and actual deployment the solar power generation resources are generally seen as having the potential to power the world, since about 7400 times the amount of solar energy reaches the earth's surface as what may be required for societies energy use (about 122 000 TW solar radiation reaches the surface of the earth while society uses about 525 EJ/yr[1] = 16.6 TW, i.e., $122\,000/16.6 = 7.4 \times 10^3$). The harvesting technologies include thermal methods such as concentrated solar power (CSP) and solar collectors, and photovoltaic (PV) methods (solar cells). In general the thermal methods use the heating of a working liquid to high (CSP) or moderate temperature and heat is extracted either to generate electricity or use it as direct heating. PV has electricity as output directly.

The insolation (solar radiation power per square meter at the earth's surface) is daily modulated between zero and a maximum that depends on the latitude on earth and the season (Figure 1). For instance in Edmonton in Canada, Delft in the Netherlands, and Astana in Kazachstan (~52° North), there is a factor of 6 between the insolation in mid summer and mid winter due to the reduced instantaneous light intensity and time of daylight (Figure S1 [4]). In Mexico City, the Western Sahara and Nagpur in central India (~19.5° North), the factor between summer and winter reduces but still reaches a sizeable factor ~1.5. Thus in principle a factor of 6 to 1.5 difference per solar power collecting footprint between seasons occurs, next to the diurnal day and night fluctuations, and varying cloud covers. These seasonal and diurnal influences multiply with each other to obtain the total solar power. For a multi Tm^2 (Terra $m^2 = 10^{12}$ $m^2 = 10^3 \times 10^3$ km^2) grid connected surface area spanning Europe and Northern Africa this will mean on average a sizeable factor ~3–4 between summer and winter insolation, modulating the day and night diurnal variation on a seasonal scale.

Wind resources within a continent sized electricity grid depend on the instantaneous wind speeds averaged over the grid surface area. It is well known that the wind power is about two times stronger in winter than in summer on northern latitudes (Figure 1). [5] Next to this seasonal timescale there is however also a diurnal periodicity of relevance. In meteorological literature a number of data studies are available of the near surface layer average wind speeds over extended surface area's in Africa and the North Atlantic, [6] and the US [7] in which thousands of local weather stations have been taken into account. The general insight gained is that there is a diurnal variation in wind speeds with significant amplitude, where the peak in wind amplitudes occurs in the afternoon and the minimum 12 h earlier in the early morning. For instance in Ref. 7, the instantaneous wind speed averaged over a ~800 × 1000 km^2 surface area on an ordinary day could be 4–5 m/s while the minimum could be 2 m/s (Figure 2). The wind speed amplitude has such diurnal pattern because it is driven by the surface temperature, i.e., the solar radiation heating the surface and atmosphere above it drives the observed wind speeds. Also more local studies in Mexico, [8] UK, [9] Scandinavia, [10] Sicily, [11] and Grenada [12] report these diurnal wind patterns. Since the kinetic energy contained in flowing air scales with the third power of the wind speed a factor 2 in wind speed amplitude means a factor 8 in recoverable energy in wind turbines.

Future wind energy implementation will also include more off shore wind parks preferably in shallow sea waters which will often be in national waters near the coastal line. In 2011 4.1 GW of the installed power was located offshore. [3] The wind patterns and diurnal variation in those is determined by the significant differential heating of the water and land area. During the day the land warms relatively fast due to solar light absorption and the cooler and denser air from the adjacent ocean flows over the land.

[13] At night the land cooling takes place by the continued emission of infrared radiation and the air flows reverse. Since the infrared is absorbed in the atmosphere more readily than the visible light (which is the basic origin of the Greenhouse effect) the nightly cooling process is on average relatively slower and the near coastal winds during night are therefore also driven less powerful than during daytime. Quantitative measurements on these diurnal ocean winds have been performed on a global scale using data from the NASA QuikSCAT satellite scatterometer launched in 1999. [13–15] The sea breeze can extend several 100 km into the sea which means that the (future) wind turbines in those regions are under the influence of such diurnal wind patterns. [14] It is also noted in Ref. 13 that in winter time the temperature difference between land and ocean is reduced and the sea breeze largely disappears.

Here a model calculation for solar power plus wind power on extended area power grids is presented. The diurnal and seasonal variation of the solar and wind power contributions add up in this model, and together they show the total renewable power variation on diurnal and seasonal timescales. Clearly there have to be made simplifying approximations in such global approach. The obtained generation patterns are compared with three different renewable solar and wind power demand patterns in order to make a first estimate for the demand of future energy storage scales. The technical features of this model approach are described in the Methods section below.

13.2 METHODS

The estimation of the integral number for future solar and wind power generation in future years is taken from Refs. 1 and 2. This is the prognosis for the total globally generated energy using these respective techniques during a selected year. The Global Energy Assessment (GEA) 2012 has been made with support by many national and international organizations; however, the values remain a prognosis. These numbers cited are in EJ/yr, i.e., there is no time structure of the output power during the day and year. To come to such time structure for solar power the geographical location of the different generating facilities connected to the large scale power grids needs to be taken into account; the insolation on a particular latitude θ_L of the earth surface varies due to the daily and seasonal varying zenith angle and the length of day. Taking the position of installed solar power at the geographical latitude θ_L and longitude φ into account, the variation $I_{solar}(t)$ throughout the year becomes

$$I_{solar}(t) = \max\left\{ I_{solar}^{total} \sum_{\theta_L=-90}^{90} \sum_{\varphi=-180}^{180} G_{solar}(\theta_L, \varphi) \frac{\cos(\theta_L, \varphi, t)}{F(\theta_L, \varphi)} \left(a_0 \right.\right.$$
$$\left.\left. + a_1 \exp\left(-\frac{k}{\cos(\theta_L, \varphi, t)}\right)\right), 0 \right\}$$

in EJ/day, where

$$\cos(\theta_L, \varphi, t) = \sin\theta_L \sin\delta(t) + \cos\theta_L \cos\delta(t) \cos\left(2\pi\left(t - \frac{1}{2}\right) - \varphi\right),$$

$$d(t) = \delta_0 \cos\left(2\pi\frac{(t-172)}{365}\right)$$

$$F(\theta_L, \varphi) = \int_0^{365} \cos(\theta_L, \varphi, t)\left(a_0 + a_1 \exp\left(-\frac{k}{\cos(\theta_L, \varphi, t)}\right)\right) dt$$

(1)

where I_{solar}^{total} equals the total integrated energy generated by solar power in a year, and δ_0 is the declination angle of $23.45°$ of the earth rotation axis. The term between brackets with parameters $a_0 = 0.4237$, $a_1 = 0.5055$, and $k = 0.2711$ stems from the transmission of solar rays through the standard clear air at sea level. [16] The $\text{Cos}(\theta_L, \varphi, t)$ is the reduction of solar power due to the angle the solar rays make with the earth surface throughout the year at latitude θ_L and longitude φ, while $F(\theta_L, \varphi)$ is a normalization factor that makes that the total solar energy generated during the year adds up to the number I_{solar}^{total}. The factor $\cos(2\pi(t-1/2)-\varphi)$ is resulting from daily rotation of the earth with the hour angle, while the function $\max\{\dots,0\}$ imposes that no negative values for solar radiation amounts can result at night or in winter. The fraction of the solar power facilities connected into the grid at the location (θ_L,φ) is represented by $G_{solar}(\theta_L,\varphi)$.

For the generated wind power a dependence on the day t of the year is approximated to match the numerical experimental data in Figures 1 and 2 as

$$I_{wind}(t) \approx \frac{I_{wind}^{total}}{365} \sum_{\theta_L=-90}^{90} \sum_{\varphi=-180}^{180} G_{wind}(\theta_L,\varphi)\left(\cos\left(2\pi\frac{(t-29)}{365}\right)\frac{43}{92}\right.$$
$$\left. +1\right)\left(\frac{3}{4}\cos\left(2\pi\left(t-\frac{16}{24}\right)-\varphi\right)+1\right)$$

(2)

in EJ/day. The second cosinus term approximates the diurnal wind power variation from Figure 2; it peaks at 4 h past midday at the longitude φ. $G_{wind}(\theta_L,\varphi)$ is again determined by the locations of the wind power facilities. The seasonal time dependence of the solar (1) and wind (2) contributions has different amplitude and phase shifts.

Clearly large uncertainties remain in the future solar and wind power implementation rate and geographical locations. However, this approach can easily be refined when new implementation data become available while the nature of the variations with time remains the same.

To come to a magnitude of the energy storage demand apart from the generating capabilities, the demand side on the short daily and the long seasonal timescales has to be taken into account. The total energy use and electricity use is relatively constant throughout the year (increasing each year1), although it shows a larger demand in mid summer (air-conditioning) and mid winter (heating and lighting) in for instance the USA. [17] The daily energy demand shows a night time low and a morning and afternoon peak as illustrated using various available energy use data in Ref.

18, the electricity demand shows a pattern which varies much less strong. [18] To model these demands $D_{total}(t)$ numerically the following equations were used:

$$D_{total}(t) = P_{total}(t) - P_{Electricity}(t) + E_{total}(t) \tag{3}$$

where

$$P_{total}(t) \approx \frac{P_{2050}\frac{30}{365}}{365} \sum_{\theta_L=-90}^{90} \sum_{\varphi=-180}^{180} \left[P_{summer}(\theta_L, \varphi, t) \left(\frac{1}{2} + \frac{1}{2}\cos\left(\frac{2\pi(t-172)}{365}\right)\right) \right.$$
$$\left. - P_{winter}(\theta_L, \varphi, t) \left(\frac{1}{2} - \frac{1}{2}\cos\left(\frac{2\pi(t-172)}{365}\right)\right) \right] A(t)$$

$$\tag{4}$$

where $P_{Summer/Winter}(\theta_L, \varphi, t)$ is the intraday primary energy demand having a pattern as in Ref. 18 (a numerical approximation is given in the supplementary material [4]). Here as an approximation the hour angle φ follows integral multiples of 15° since the energy use will be related to the time in individual time zones. As a further refinement the hour angle φ could be chosen according to the actual longitudes of a country (which is not done here). The 172nd day is the 21st of June. The function A(t) has the form: $A(t) = (1+0.075\cos(4\pi(t-217)/365))$ $(1-0.045\cos(2\pi(t-217)/365))$ and approximates the variation in primary energy use throughout the year as observable in Ref. 17 (peak in winter and lower peak in summer). The electricity demand $E_{total}(t)$ has the same form of equation as (4) but now with $E_{Summer/Winter}(\theta_L, \varphi, t)$ replacing $P_{Summer/Winter}(\theta_L, \varphi, t)$ and A(t)=1. $P_{Electricity}(t) = E_{Total}(t)/0.50$ is the primary energy required to generate the electricity $E_{Total}(t)$ assuming a 50% efficiency; it has thus the same time dependence as $E_{Total}(t)$. 50% efficiency is lower than what can be reached in modern gas fired combined cycle power stations; however, this choice accounts for the use of, e.g., lower efficiency coal fired power stations as well. It is subtracted in $D_{Total}(t)$ to come to an energy demand which is usable energy for the end consumer, and in order to be able to subtract the renewable electricity from the fossil based electricity.

A second factor that is important to determine a future energy storage scale is the connectivity of the generating facilities on large extended power grids. This determines how much electricity from distant sources, transmitted at the speed of light through the electricity grid, contributes to the instantaneous integrated output of the grid. The plans for long distance power transport include connected grids on the size of, e.g., Europe plus northern Africa; i.e., latitudes from ~24° to 62° and distances ~2000 × 3000 km². Much larger grids are not considered in view of the large cost and increasing transport losses; the cost will increase faster than linear with distance traveled since also the amount of peak power to be transported will grow with the surface area that is connected. This distance constraint makes that here a calculation for a large but limited range of latitudes and longitudes is considered. The result can, however, be extrapolated to the worldwide generating capabilities and storage demands because other independent grids will mostly be located on similar latitudes.

The future electricity demand will be modeled according to three different scenarios. Scenario I is having the year average electricity demand scaled up with the same factor as the total energy is scaled up. Scenario II is having a year average electricity demand scaled up to the total yearly renewables generation, and Scenario III is having the same electricity demand as Scenario I but added to that a demand contribution having the time structure of the primary energy demand. The total year average demand in Scenario III equals the yearly renewables generation. Scenarios II and III thus indicate the electrification of previously fossil powered demands in two different ways. Having the demand and supply characteristics modeled as described above, the storage requirement on different timescales can be estimated from the difference between the demand and the supply during chosen periods of time. The formulas used here are in the Supplementary material. [4]

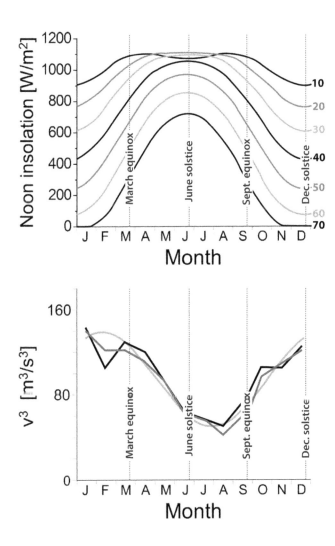

FIGURE 1: Top: daily insolation at noon during the months of the year on the indicated northern latitudes. See also Fig. S1 in the supplementary material [4] for the total daily insolation. Bottom: estimated average cubed windspeed v^3 in the US for on shore and off shore locations (based on data from Ref. 5), and a simple sinusoidal approximation as in Eq. (2).

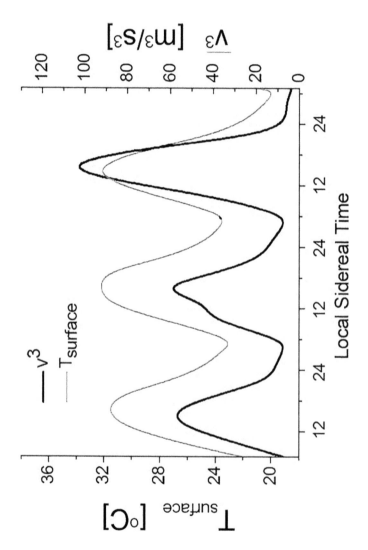

FIGURE 2: Daily averaged v^3 for a large 800×1000 km^2 area in the US and the average surface temperature for three consecutive days (constructed from data in Ref. 7).

13.3 RESULTS

13.3.1 RENEWABLE POWER VARIATION
ON CONTINENT SIZED GRIDS

To obtain numerical values for the solar and wind power during the year Eqs. (1) and (2), respectively, are applied. The amplitudes $I_{solar,wind}^{total}$ are taken from estimated mean global values in the GEA 2012 report. [1] For comparison also a mean value from the Intergovernmental Panel on Climate Change (IPCC) 2012 report [2] is shown. The values used for $(I_{solar}^{total}, I_{wind}^{total})$ are from the GEA-Mix/IPCC-Median estimates: (70,40)/ (2.2, 11.3) EJ and (170,80)/(12.8, 23) EJ for 2030 and 2050, respectively. It should be noted that the IPCC number used stems from the median of a large number of separate studies with a very large spread in the predictions. The GEA-Mix values are significantly higher than this IPCC-Median but are still within the ranges considered by the IPCC report. Since the GEA-Mix report is aimed at specifically energy forecasting, we use these values in the remainder; however, the IPCC values are used to produce the indicated graphs for comparison. The modeled power outputs are in Figure 3 (averaged over a day), 4 (instantaneous) and S2, S3, S4 [4] graphs for an extended power grid. The range of latitudes for the grid used here as an approximation is between 20° and 60° with $G_{solar,wind}$ $(20°,\varphi) = 5\%$, $G_{solar,wind}$ $(30°,\varphi) = 15\%$, $G_{solar,wind}$ $(40°,\varphi) = 35\%$, $G_{solar,wind}$ $(50°,\varphi) = 35\%$, and $G_{solar,wind}$ $(60°,\varphi) = 10\%$. The average latitude is then 43°, which thus assumes that the solar power installations are considerably more south than currently. In addition equal contributions of a longitude −5°, 5°, and 15° were taken corresponding to about 1600 km in the east-west direction. Such range of latitude and longitude corresponds roughly to grids spanning Northern Africa to Europe, and it is also within the latitude range where, e.g., the high density population and energy use is of the USA, North/East India, and China. The majority of installed capacity may be anticipated in such latitude range, e.g., to minimize transport costs and crossing of state borders.

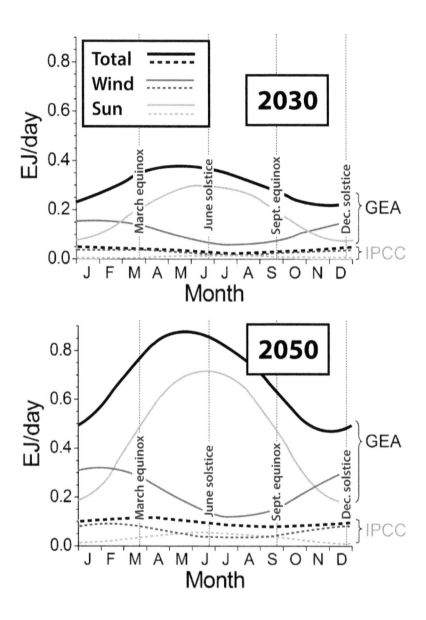

FIGURE 3: Estimated output per day of wind and solar power in the months of the years indicated. Due to the geographical locations of the facilities above the equator (see text) a significant variation in output power throughout the year is expected. GEA and IPCC (lower curves) indicate two different levels of renewable energy implementation.

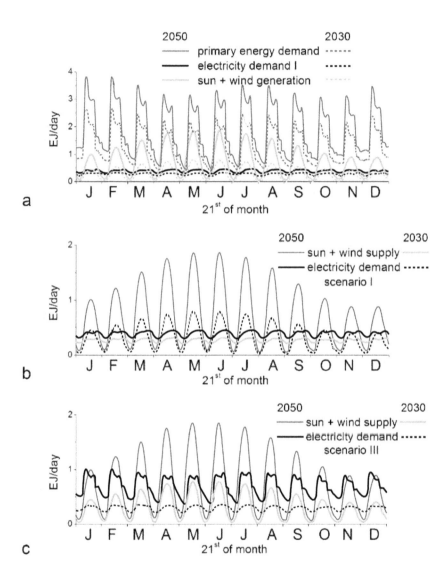

FIGURE 4: (a) Estimated output from the model grid on a selected latitude and longitude distribution (see text) for 2030 and 2050 during the 21st of the months indicated, scaled up to world scale and using the GEA-Mix implementation of renewables. Also the total energy use and the electricity use in scenario I are indicated. The renewable energy demand patterns I and III are included in Figures 4(b) and 4(c), respectively.

13.3.2 ENERGY DEMAND PATTERNS

The energy use is modeled according to available current demand patterns throughout the year for electricity and primary energy. The assumption is that much of the current and future demand is, and will be, organized in time for functional reasons that cannot easily be altered to a large extent, but the use of electricity relative to primary energies can be altered. To obtain numerical values for time dependent demand patterns an approximation of known data is used as described in the Methods section. In Figure 4(a) the estimated demand patterns for the 21st of each month are given for electricity and primary energy from which the primary energy (mainly fossil fuels) required to generate electricity has been subtracted. The ratio primary energy to electricity demand is initially simply kept constant at the current ratio while both are scaled up to reach the future total energy use level in the GEA-Mix scenario (demand type scenario I). Comparing the instantaneous electricity demand with the sun + wind generating capabilities in Figure 4(a) and 4(b) immediately shows that by 2050 the instantaneous electricity demand in this scenario I will be significantly lower than the instantaneous renewables generation during most of the day, even in winter. To make use of the renewable power will thus necessitate (1) conversion of applications to use electricity instead of primary energies where that is possible, and (2) realize sufficient electricity supply throughout the year to power these additional electric applications, or alternatively (3) directly convert the renewable power to a form of primary energy that can be accepted by applications using primary energy from fossil or renewable origin. This may include the use of solar heat directly in combination with storage when possible. In principle, conversions of electricity to other primary energies involves losses; for that reason direct use as electricity—if possible—is of advantage.

(1) and (2) denote that the year average electricity demand will be best of the order of magnitude of the average renewable electricity generation to minimize the amount of conversion losses. Coping with the summer daytime peak and lower output during winter and at night will mean partly storing the peak electricity supply from renewables for use at night and in winter. On seasonal timescales, this involves renewable electric-

ity conversion into a suitable form of stored primary energy or fuel. The same conversion to a suitable primary energy is also the case in (3), but then without the long term storage. In Figure 4 there are therefore two demand scenario's given: scenario I which has the pattern of the electricity demand scaled up to a year average demand increased by the same factor as the total energy demand is increased, and scenario III with the electricity demand as in I, but added to that an additional primary energy demand pattern bringing the total year average demand to the year average renewables generation level. Scenario II is the demand pattern similar to I, but then scaled up to reach the year average renewables generation, i.e., increased more rapidly than total energy use (Figure S3 [4]).

13.3.3 ENERGY STORAGE AND CONVERSION SCALES

To cope with the described systematic variability of renewable electricity generation, the current approach is to power up and down fossil fuel powered stations. In this way renewables reduce the operational filling factor of these stations, have an impact on the business model of these facilities, but do not really replace fossil fuel generating capabilities. The additional storage capacity required then "only" covers the time it takes to power up or down the fossil fuel powered stations to maintain the grid stability, if possible. From the modeled daily output by 2050 in Figure 4 it is clear that coping with the renewables peak power by switching off fossil power alone is not enough, since significantly more renewable power is produced than can be switched off. Thus assuming that renewable power generating capabilities essentially should replace fossil fuel based power generating capabilities and electricity will be converted to primary energy forms like fuels this will necessarily represent renewable energy storage and conversion on an unprecedented scale. The dependence on fossil fuel and the associated CO_2 emission can then be reduced as illustrated in Figure 5.

The demand for renewable energy storage and conversion is estimated as follows. First, the daytime renewable electricity generation which is larger than the demand is stored for later use in the night, leading to a short time storage demand. Note that such short term storage also includes load or demand time shifting of, e.g., electrical vehicles. The remaining surplus of renewable energy is assumed to be converted to primary energy (e.g., high energy density fuels, see below) and stored for the longer seasonal timescale. Using this approach Figure 6 results for the magnitude of short and long term energy storage demand. In Table I the short and long term storage for scenarios I, II, and III is summarized. Clearly long term storage only takes place in summer, and this is in this approach completely used up in winter. Short term storage would still occur in winter when the instantaneous generation is larger than the instantaneous use of renewable power. The result is that scenario I would generate the need for significantly more short and long term storage capacity, essentially because the demand is too low compared to the generation. In scenario I in fact more energy is stored in summer than can be used in winter to generate electricity. Scenarios II and III will thus be more realistic and favorable, since the renewable energy generated is also completely used (with or without storage/conversion). There is not so much difference between the latter two scenario's with up to 0.2 EJ/day short term and around 30 EJ $= 8333$ TW h per season long term storage by 2050 worldwide. This corresponds to 20 days of current primary energy use in the GEA prognosis for 2050 ($\sim$525 EJ/365 $= 1.4$ EJ per day). The short term storage then corresponds to about 0.7% of the seasonal storage, and on average $\sim$8 kWh installed storage capacity per person on an earth with 7×10^9 inhabitants.

TABLE 1: Summary of energy storage demands in demand scenarios I, II, and III, based on the model described and GEA prognosis for average renewable energy generation in 2030 and 2050.

Demand scenario	Short term storage demand 10−2 EJ/day		Long term, seasonal, storage demand, EJ	
	2030	2050	2030	2050
I	2–8	5–9	16.5	108
II	2–9	3–20	10.7	27.5
III	1.5–8	1–14	10.8	29.2

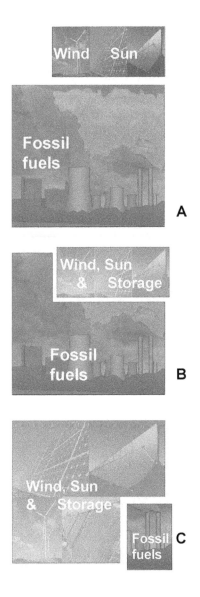

FIGURE 5: Schematic of installed rated power. (a) Fossil generating capabilities and renewable solar and wind power without renewable energy storage option. To guarantee security of supply practically the full conventional fossil capacity will be required. (b) and (c) With long and short term storage of renewable energy part of the fossil capabilities can be replaced progressively by renewable powered facilities, or be fueled with renewable fuel.

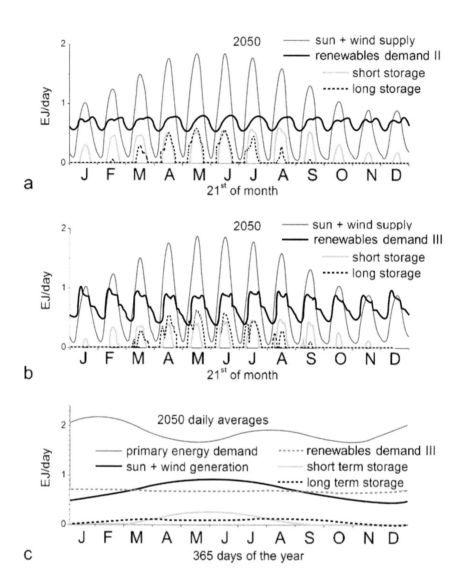

FIGURE 6: The pattern of renewable energy storage during the indicated days of the year 2050 in scenarios II (Figure 6(a)) and III (Figure 6(b)) for the grid on the Northern hemisphere as indicated in the text, and scaled to the total world electricity and energy use. The integral values for 1 day in (c); the primary plus the renewable energy demand equals the total usable energy demand.

13.4 DISCUSSION

The solar and wind power generation on large scale grids will vary strongly and systematically on both a daily and seasonal timescale. The comparison with the demand for energy during the day and seasons, results in significant storage demands on different timescales if one intends to completely use the energy that is generated. As far as we know this is the first such global scale estimation. At the same time, the different prognosis of solar and wind power implementation of GEA and IPCC reports illustrate that the actual range of storage requirements can vary significantly. Also in the model for the time dependent demand choices are made that can influence the storage scale. However, the method used can be easily adapted when the implementation of solar and wind power as well as the demand patterns progress in the coming decades. Additional sources of discrepancies will be the variation of the extended weather conditions from day to day and season to season, e.g., a cold winter or hot summer will have an impact on the energy use and generation in sections of the world. In that sense we used averages of the weather and absorbed all local and temporal fluctuations that will be present in reality. In practice, however, one can propose that this will rather increase than decrease the demand for short term and long term storage, since the stored energy also provides partial independence from adverse renewable energy generation conditions.

In the solar radiation estimation we did not take into account the orientation of solar panels towards the sun, but rather used the W/m^2 horizontal surface. This simplification was chosen since in practice many PV panels on, e.g., roofs will not have the ideal orientation and because then also shadowing effects may be taken into account. The effect of this approximation will have some influence on the shape of the daily solar power generation curves, but since the total generated solar energy is normalized not on the total yearly solar power generation.

More extended grid scale, extending towards the southern hemisphere would address the summer winter variability, while even larger east-west grids also spanning the entire globe would also address the day and night variability. However, the feasibility of such power harvesting and grid is not beforehand clear in view of geographical factors such as available land area, depth of oceans, and geopolitics. Also the losses for each 1000 km may be 3% for high voltage DC lines, [19,20] the AC-DC conversion, and back taking an additional 1.5% each. For distances up to 20 000 or 30 000 km the transmission then amounts to $0.985^2 \times 0.97^{20\,or\,30} = 0.53$ or 0.39. In addition, such very long distance grid should transport not on GW scale, as local power grids are currently built for, but rather on the level of power use of a continent, i.e., TW scale, which will also make it highly costly, if feasible. Thus also with such investment in a world grid, losses are non-negligible (and cannot be reused). For less long distances, e.g., the distance from Norway to the Sahara (~4100 km), smaller losses occur (transmission = 0.86), but as stated above the daily and seasonal storage are not addressed. Counteracting seasonal effects could be possible with a grid extending from Norway to below the equator (e.g., Angola) which is a distance of 9300 km (transmission 0.73), but then the day and night variability is not addressed.

The solar power generation as estimated above is located on the northern hemisphere where the largest part of the human population is located. For the southern hemisphere the same energy generation and storage scheme can be used, but then shifted 6 months and 12 h in time. Since the power grid will not be connected over such long distances the global energy storage requirements will be similar, proportional to the overall installed capacities. The method used here adds all generation, use, and storage shifted to one central time, effectively adding up to a world wide contribution.

For smaller grid scales in principle, the weather conditions become less averaged and more fluctuating, and also more dependent on the specific location. The "short term" storage facilities then likely needs extension of the capacity towards storage times of days in order to deal with several unfavorable renewables generation days. The seasonal scale will depend on the more local average climate.

Based on the above both short term daily and long term seasonal storage is required on scales that will only be feasible for few storage options. [21–23] Important scalable options for short term storage are heat storage [24] (high temperature storage for CSP, low temperature heat) and batteries [25] (sun-PV, wind). Currently applied pumped hydropower relies on the presence of suitable geographic factors and is thus limited in scale.

The use of batteries as electricity store will require low cost and far improved lifetime during prolonged cycling of the batteries; an energy store may be built for at least 20–30 years continuous use (105 h) compatible to, e.g., PV lifetimes. For seasonal scale energy storage artificial fuels are required. Hydrogen can be produced from renewable electricity and water [26,27] using, e.g., alkaline electrolysis with relatively inexpensive Ni based catalysts. Ammonia stands out in energy density for static stores as it is liquid at 10 bars and room temperature (RT) with an energy density of 22.5 MJ/kg higher heating value (HHV), and it contains only abundant H and N. [28,29] More conventional fuels with highest energy density up to 49 MJ/kg (propane) would require carbon, but can in principle be generated from renewable power, water and CO_2 using existing technologies. [30] However, ultimately in a fossil fuel poor energy economy CO_2 has to be captured from air since central point sources would produce only a fraction of the needs. [31]

13.5 BIOFUELS FOR LARGE SCALE STORAGE?

In Refs. 26 and 32, the use of biomass for biofuel generation is essentially excluded as viable large scale option. In the IPCC report, [2] however, it is indicated with large uncertainty that biomass could contribute between 10% and 100% of future energy use. To gain some insight in this matter we use recent estimation of the energy production from biofuels per year to come to a surface area that would be required for producing chosen amounts of biofuel. As a reference the current energy use is expressed in required production of ml oil/m^2 of earth surface and biofuel production in terms of oil equivalent per m^2 earth surface (Fig. 7).

With a higher heating value of 34 MJ/l (gasoline) and an earth total surface area of 5.1×10^{14} m^2 the current energy use of ~500 EJ/yr equals 28.8 ml oil/m^2; an oil film of 28.8 μm thick on the entire globe (Fig. 7). One of the often mentioned high yield biofuel sources that would not compete with food production directly is switchgrass. Its net energy yield in the form of bioethanol is reported [33] as 6 MJ m^{-2} yr^{-1} which is equivalent to 176 ml oil $m^{-2}yr^{-1}$ (heating value of ethanol is 23.43 MJ/l). In order to power the world with switchgrass bioethanol one thus requires at least $28.8/176 = 16.3\%$ of the entire surface of the globe, or ~half of the land area (assuming appropriate climate conditions). For poplar trees the result is similar. [34] For biodiesel from palmoil an estimated 0.6 l m^{-2} yr^{-1} is reported. In general for biodiesel 2.2 units of oil are the net energy gain for a harvest of 3.2 units, [35] i.e., $600 \times 2.2/3.2 = 412$ ml m^{-2} yr^{-1} is the gain. So for palmoil $28.8/412 = 7.0\%$ of the earth surface would be required to produce 500 EJ/yr. These numbers are relative to the entire surface of the globe, including oceans, poles, deserts, permafrost and mountains, regions with wildly different and incompatible climate conditions. The current agricultural area is quoted as 49×10^6 km$^2 = 4.9 \times 10^{12}$ m^2 in 2010 by the Food and Agriculture Organization of the United Nations, which is almost 1% of the earth's surface area. For switchgrass and poplar, and palmoil thus [16], respectively, 7 times more than the current agricultural area would be required to produce sufficient biofuel to reach the higher limit of 500 EJ/yr. In such perspective 10% of that appears as an enormous amount of additional area which needs to be made accessible for agricultural activities in a sustainable manner. In addition the biomass will need to act as a valuable carbon source for materials fabrication and as such may become too precious as fuel.

Biofuels could also be considered to cover "only" the seasonal storage needs as described above, next to solar and wind power. In that case the 27 EJ by 2050 could be realized with a lower demand on space, which however still equals for palmoil ~7% $\times$ 27/500 = 0.38% of the surface of the earth. This corresponds to 42% of the current agricultural area (which will generally not be suitable for growing palmoil). In comparison, solar PV with 10% efficiency (which is half of current commercially available Si single junction solar cells), an insolation of 2 kW h m^{-2} day^{-1} (which is less than half of that in Spain), and 60% efficient electrolysis to produce hydrogen would require 1.7×10^5 km^2 ~3.5% of the agricultural area to collect the 27 EJ. This is a much lower demand on space, even with moderate efficiency solar cells and low insolation values. Also this PV space can be on rooftops or dry, barren land not suitable to grow crops or difficult to reach by road.

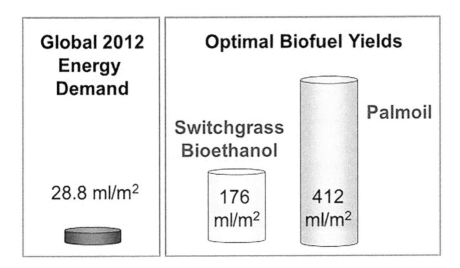

Figure 7: Illustration comparing the current yearly energy demand with the amounts of experimentally verified yearly optimal biofuel yields. The unit is expressed in ml oil equivalent per square meter of total earth surface for the demand and in units of ml of oil equivalent per used square meter for the yields.

REFERENCES

1. GEA Global Energy Assessment, Toward a Sustainable Future (Cambridge University Press, Cambridge, UK and New York, NY, USA and the International Institute for Applied Systems Analysis, Laxenburg, Austria, 2012); online at www.globalenergyassessment.org

2. Renewable Energy Sources and Climate Change Mitigation; Special Report of the Intergovernmental Panel on Climate Change (Cambridge University Press, 2012).

3. IPCC, Global Wind Energy Council, see http://www.gwec.net/publications for global wind report 2011, annual market update.

4. See supplementary material at http://dx.doi.org/10.1063/1.4874845 for additional figures, numerical approximation of the electricity and energy demand patterns, as well as the energy storage calculations. [Supplementary Material]

5. X. Lu, M. B. McElroy, and J. Kiviluoma, " Global potential for wind-generated electricity," Proc. Natl. Acad. Sci. U.S.A. 106, 10933–10938 (2009).http://dx.doi.org/10.1073/pnas.0904101106

6. A. Dai and C. Deser, " Diurnal and semidiurnal variations in global surface wind and divergence fields," J. Geophys. Res., D: Atmos. 104, 31109–31125 (1999).http://dx.doi.org/10.1029/1999JD900927

7. D. L. Zhang and W. Z. Zheng, " Diurnal cycles of surface winds and temperatures as simulated by five boundary layer parameterizations," J. Appl. Meteorol. 43, 157–169 (2004).http://dx.doi.org/10.1175/1520-0450(2004)043<0157:DCOSWA>2.0.CO;2

8. W. J. Shaw, M. S. Pekour, R. L. Coulter, T. J. Martin, and J. T. Walters, " The day-time mixing layer observed by radiosonde, profiler, and lidar during MILAGRO," Atmos. Chem. Phys. Discuss. 7, 15025–15065 (2007).http://dx.doi.org/10.5194/acpd-7-15025-2007

9. G. Sinden, " Characteristics of the UK wind resource: Long-term patterns and re-lationship to electricity demand," Energy Policy 35, 112–127 (2007).http://dx.doi.org/10.1016/j.enpol.2005.10.003

10. H. Holttinen, " Hourly wind power variations in the Nordic countries," Wind Energy 8, 173–195 (2005).http://dx.doi.org/10.1002/we.144

11. S. Bivona, R. Burlon, and C. Leone, " Hourly wind speed analysis in Sic-ily," Renewable Energy 28, 1371–1385 (2003).http://dx.doi.org/10.1016/S0960-1481(02)00230-6

12. D. Weisser, " A wind energy analysis of Grenada: An estimation using the 'Weibull' density function," Renewable Energy 28, 1803–1812 (2003).http://dx.doi.org/10.1016/S0960-1481(03)00016-8

13. S. T. Gille, S. G. Llewellyn Smith, and S. M. Lee, " Measuring the sea breeze from QuikSCAT Scatterometry," Geophys. Res. Lett. 30, 14–11, doi: doi: http://dx.doi.org/10.1029/2002GL016230 (2003).

14. S. T. Gille, S. G. L. Smith, and N. M. Statom, " Global observations of the land breeze," Geophys. Res. Lett. 32, L05605, doi: doi: http://dx.doi.org/10.1029/2004GL022139 (2005).

15. M. H. Pickett, W. Q. Tang, L. K. Rosenfeld, and C. H. Wash, " QuikSCAT sat-ellite comparisons with nearshore buoy wind data off the US West Coast," J. Atmos. Ocean. Technol. 20, 1869–1879 (2003).http://dx.doi.org/10.1175/1520-0426(2003)020<1869:QSCWNB>2.0.CO;2

16. H. C. Hottel, " A simple model for estimating the transmittance of direct solar ra-diation through clear atmospheres," Sol. Energy 18, 129–134 (1976).http://dx.doi.org/10.1016/0038-092X(76)90045-1

17. U.S. Energy Information Administration, see http://www.eia.gov/totalenergy/data/monthly/#summary for monthly energy review November 2012.

18. D. J. Sailor and L. Lu, " A top-down methodology for developing diurnal and sea-sonal anthropogenic heating profiles for urban areas," Atmos. Environ. 38, 2737–2748 (2004).http://dx.doi.org/10.1016/j.atmosenv.2004.01.034

19. G. Asplund, " Ultra high voltage transmission," ABB Rev. 2, 22–27 (2007).

20. A. G. Siemens, see http://www.energy.siemens.com/hq/en/power-transmission/hvdc/hvdc-ultra/#content=Benefits%20 for high voltage DC loss percentage.

21. W. F. Pickard, " A nation-sized battery?," Energy Policy 45, 263–267 (2012).http://dx.doi.org/10.1016/j.enpol.2012.02.027

22. N. Armaroli and V. Balzani, " Towards an electricity-powered world," Energy Envi-ron. Sci. 4, 3193–3222 (2011).http://dx.doi.org/10.1039/c1ee01249e

23. A. Zuttel, A. Remhof, A. Borgschulte, and O. Friedrichs, " Hydrogen: The future energy carrier," Philos. Trans. R. Soc., A 368, 3329–3342 (2010).http://dx.doi.org/10.1098/rsta.2010.0113

24. M. Liu, W. Saman, and F. Bruno, " Review on storage materials and thermal performance enhancement techniques for high temperature phase change thermal storage systems," Renewable Sustainable Energy Rev. 16, 2118–2132 (2012).http://dx.doi.org/10.1016/j.rser.2012.01.020

25. B. Dunn, H. Kamath, and J. M. Tarascon, " Electrical energy storage for the grid: A battery of choices," Science 334, 928–935 (2011).http://dx.doi.org/10.1126/science.1212741

26. J. A. Turner, " A realizable renewable energy future," Science 285, 687–689 (1999). http://dx.doi.org/10.1126/science.285.5428.687

27. J. A. Turner, " Sustainable hydrogen production," Science 305, 972–974 (2004). http://dx.doi.org/10.1126/science.1103197

28. R. Lan, J. T. S. Irvine, and S. W. Tao, " Ammonia and related chemicals as potential indirect hydrogen storage materials," Int. J. Hydrogen Energy 37, 1482–1494 (2012).http://dx.doi.org/10.1016/j.ijhydene.2011.10.004

29. A. Klerke, C. H. Christensen, J. K. Norskov, and T. Vegge, " Ammonia for hydrogen storage: Challenges and opportunities," J. Mater. Chem. 18, 2304–2310 (2008). http://dx.doi.org/10.1039/b720020j

30. W. Haije and H. Geerlings, " Efficient production of solar fuel using existing large scale production technologies," Environ. Sci. Technol. 45, 8609–8610 (2011).http://dx.doi.org/10.1021/es203160k

31. A. Goeppert, M. Czaun, G. K. S. Prakash, and G. A. Olah, " Air as the renewable carbon source of the future: An overview of CO2 capture from the atmosphere," Energy Environ. Sci. 5, 7833–7853 (2012).http://dx.doi.org/10.1039/c2ee21586a

32. J. A. Turner, " Biomass in the energy picture—Response," Science 285, 1209–1210 (1999).

33. M. R. Schmer, K. P. Vogel, R. B. Mitchell, and R. K. Perrin, " Net energy of cellulosic ethanol from switchgrass," Proc. Natl. Acad. Sci. U.S.A. 105, 464–469 (2008). http://dx.doi.org/10.1073/pnas.0704767105

34. K. Sanderson, " US biofuels: A field in ferment," Nature 444, 673–676 (2006).http://dx.doi.org/10.1038/444673a

35. U.S. Department of Agriculture and U.S. Department of Energy, Life cycle inventory of biodiesel and petroleum diesel for use in an urban bus, Report No. NREL/SR-580-24089, 1998.

Author Notes

CHAPTER 1

Acknowledgments

The research is financially supported by National Natural Science Foundation of China (Program no. 50907010), Research and Innovation Project for College Postgraduates of Jiangsu Province (Program no. CXLX11_0112), and The Scientific Research Foundation of Graduate School of Southeast University (YBJJ1132). The authors would like to thank the National Meteorological Information Centre, China Meteorological administration. They are also grateful to the anonymous reviewer for his/her constructive comments and suggestions on this paper.

CHAPTER 2

Conflict of Interest

The authors declare that there is no conflict of interests regarding the publication of this paper.

Acknowledgments

The authors are grateful to National Aeronautics and Space Administration (NASA) for making data for this study freely available.

CHAPTER 3

Acknowledgments

The authors would like to thank the EPSRC for funding this research. Grant number EP/G037728/1.

CHAPTER 4

Conflict of Interest
The authors declare that there is no conflict of interests regarding the publication of this paper.

CHAPTER 5

Conflict of Interest
The authors declare that there is no conflict of interests regarding the publication of this paper.

CHAPTER 7

Acknowledgments
RW acknowledges the support of Hong Kong Research Grants Council grant (CUHK403612) and National Natural Science Foundation of China grants (41228006 and 41275081).

CHAPTER 8

Conflict of Interest
The authors declare that there is no conflict of interests regarding the publication of this paper.

CHAPTER 10

ACKNOWLEDGMENTS

This paper is funded by a grant/cooperative agreement from the National Oceanic and Atmospheric Administration, NA0OAR4311004. The views expressed herein are those of the authors and do not necessarily reflect the views of NOAA or any of its subagencies. Special thanks are given to Alessandra Giannini and Michael Tippet of IRI for their guidance and advice.

CHAPTER 12

Conflict of Interest

The authors declare that there is no conflict of interests regarding the publication of this paper.

Acknowledgments

This study was supported by Karadeniz Technical University Scientific Research Projects Unit, Project no. 2008.112.004.2.

CHAPTER 13

Acknowledgments

This manuscript is the result of joint research in the Delft Research Centre for Sustainable Energy and the 3TU Centre for Sustainable Energy Technologies. Support from NWO/ACTS under Project No. 053.61.017 is also acknowledged.

Index

Milton Keynes UK
Ingram Content Group UK Ltd.
UKHW050256161024
449569UK00042B/1726

9 781774 637098